T0282506

CAMBRIDGE LIBRARY COLLECTION

Books of enduring scholarly value

Technology

The focus of this series is engineering, broadly construed. It covers technological innovation from a range of periods and cultures, but centres on the
technological achievements of the industrial era in the West, particularly
in the nineteenth century, as understood by their contemporaries. Infra-
structure is one major focus, covering the building of railways and canals,
bridges and tunnels, land drainage, the laying of submarine cables, and the
construction of docks and lighthouses. Other key topics include develop-
ments in industrial and manufacturing fields such as mining technology, the
production of iron and steel, the use of steam power, and chemical processes
such as photography and textile dyes.

Concrete

Clerk of works to an aristocratic landowner, Thomas Potter possessed
considerable practical experience when he published this work in 1877.
His intention was to provide a source of helpful information relating to a
building material that was being increasingly used in Victorian construction,
yet not without detractors, who objected on aesthetic as well as technical
grounds. Clearly enthusiastic about concrete's potential applications,
Potter seeks to give a balanced assessment of its usefulness and versatility.
While the text does not discuss the chemical processes involved, it does
cover aggregates, matrices, how to mix the two, the apparatus needed, the
construction of walls, floors and roofs, and the costs and disadvantages of
using concrete. The book also features several contemporary advertisements,
including one for 'Potter's Concrete Building Apparatus and Appliances'. Of
related interest, Charles William Pasley's *Observations on Limes, Calcareous
Cements, Mortars, Stuccos, and Concrete* (1838) is also reissued in this series.

Cambridge University Press has long been a pioneer in the reissuing of out-of-print titles from its own backlist, producing digital reprints of books that are still sought after by scholars and students but could not be reprinted economically using traditional technology. The Cambridge Library Collection extends this activity to a wider range of books which are still of importance to researchers and professionals, either for the source material they contain, or as landmarks in the history of their academic discipline.

Drawing from the world-renowned collections in the Cambridge University Library and other partner libraries, and guided by the advice of experts in each subject area, Cambridge University Press is using state-of-the-art scanning machines in its own Printing House to capture the content of each book selected for inclusion. The files are processed to give a consistently clear, crisp image, and the books finished to the high quality standard for which the Press is recognised around the world. The latest print-on-demand technology ensures that the books will remain available indefinitely, and that orders for single or multiple copies can quickly be supplied.

The Cambridge Library Collection brings back to life books of enduring scholarly value (including out-of-copyright works originally issued by other publishers) across a wide range of disciplines in the humanities and social sciences and in science and technology.

Concrete

*Its Use in Building and the Construction
of Concrete Walls, Floors, Etc.*

Thomas Potter

CAMBRIDGE
UNIVERSITY PRESS

University Printing House, Cambridge, CB2 8BS, United Kingdom

Cambridge University Press is part of the University of Cambridge.
It furthers the University's mission by disseminating knowledge in the pursuit of
education, learning and research at the highest international levels of excellence.

www.cambridge.org
Information on this title: www.cambridge.org/9781108070515

This edition first published 1877
This digitally printed version 2014

ISBN 978-1-108-07051-5 Paperback

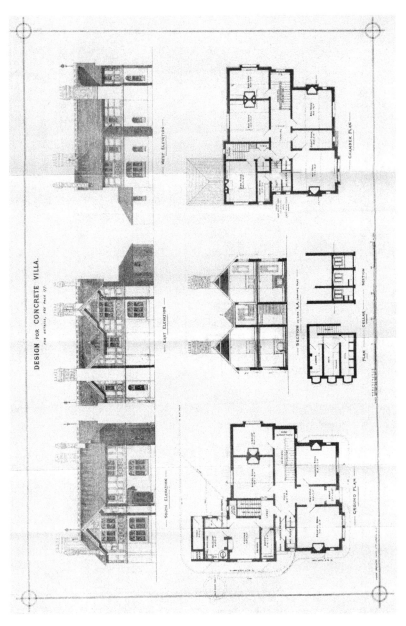

The material originally positioned here is too large for reproduction in this reissue. A PDF can be downloaded from the web address given on page iv of this book, by clicking on 'Resources Available'.

CONCRETE:

ITS USE IN BUILDING

AND THE

CONSTRUCTION OF

CONCRETE WALLS, FLOORS, ETC.

BY

THOMAS POTTER,

Clerk of Works to the Right Honourable Lord Ashburton.

WITH ILLUSTRATIONS.

London:

E. & F. N. SPON, 48, CHARING CROSS;

New York, 446, Broome Street;

AND

F. W. Reynolds & Co., 73, Southwark Street, London.

UNWIN BROTHERS,

PRINTERS.

PREFACE.

This small work is intended to deal in a practical manner with the treatment of a material that has yet undoubtedly to come into more general use. No attempt has been made to give the "reason why" in many cases the behaviour of concrete stands so pre-eminently above all other materials for special purposes, nor to explain the theory of cement setting, and the chemical process developed in the operation of making concrete; these subjects have already found more capable exponents; but simply to give an unprejudiced statement of the various forms in which concrete can be applied, and most suitable purposes for which it is adapted. Also to reconcile those extremes of individual opinion engendered on the one hand by inability to see that concrete possesses any defects, and on the other by that veneration for existing forms and processes of building, that creates a burst of indignant feeling when innovation of any kind is even suggested.

Statements herein made as facts are the result of personal observation and experience, except where the authorship is acknowledged, and are based upon long practice under variable conditions.

July, 1877.

CONTENTS.

INDEX.

A.

T.

CONCRETE:

ITS USE IN BUILDING, AND FOR OTHER PURPOSES.

INTRODUCTION.

WHEN we reflect, that the inhabitants of civilised countries pass, on an average, more than a third, perhaps nearer one half, a lifetime within the walls of their dwellings, or of other buildings, it is something marvellous what apathy, as a rule, is shown to their healthfulness. In no other instance probably is the old adage, " Out of sight out of mind," of such universal application, as in the utter disregard paid by so many to the ordinary wholesome construction and maintenance of the houses they live in. If sinks have the indispensable "traps," it is considered of little importance to ascertain whether they are really what they profess to be, or, as in many known cases they have proved to be, " fever and death traps." Provided there are drains supposed to convey away the sewage, it is assumed there is no need to inquire if they are sound, whether they lead through portions of buildings that should by all means be avoided, or if it were not possible to keep them outside altogether, and whether or not they discharge into cesspools hermetically sealed and never emptied. With the " acme of

modern civilisation," a water-closet, it is quite possible
the foul gases engendered therein have no means of
escape, other than through an inefficient water trap,
direct into rooms perhaps occupied as sleeping apartments, or the water for domestic use is obtained from
the same cistern which supplies the water-closet, and
the cistern itself has perhaps not been cleansed within
the memory of anyone who could possibly answer that
question. So that the floors of rooms, a few inches
removed from the natural soil, or their sleeper timbers
partially buried therein (not at all an unlikely occurrence with cheaply-built cottages and villas), are planed
to look smooth and clean, it is most unusual to remove
a board and look beneath, although probably some
curiosities of vegetation and specimens of fungi,
emitting most unwholesome smells, would be discovered, and the space between the floor and the earth
found to be deficient of the slightest ventilation.

But if a house has a showy portico, with a flight
of stone steps leading thereto, or some other form of
attraction, it will—with by far the greater number—
atone for a multitude of shortcomings in other respects;
and if moreover the walls are brick-built, the bricks
coloured to make them look uniform and hide their
defects, and the joints "tuck pointed," scarcely one
person in a thousand would consider for one moment,
whether or not they would absorb an indefinite
amount of rain, and what must become of all the
moisture they imbibe. If the walls will just, and
but just, keep the wet from penetrating as far as
the paper-hangings, the house is declared to be perfectly dry, and no thought is given as to the mass of
damp material encompassing the inmates on all sides,

and explaining the chilly feeling experienced, except
in hot weather, on entering a room where there is no
fire; should rain, however, find its way completely
through walls, plastering, and paper, it is assumed, if
proposed to make a perfect cure, that the proper
thing to do is to batten the walls, and then lath-and-
plaster them, leaving an uninterrupted run for rats
and mice, who, if they find no other means of entry,
will consider it but little trouble to eat their way
through the brickwork itself; but if battening and
plastering is too costly, the more economic method of
battening, canvassing, and papering is resorted to.
In either case the inmates are little better off than if
wrapped in a thick wet blanket, but so that the evil
cannot be seen its existence is ignored. If, however,
houses are inhabited, or other reasons prevail, then
the brickwork is treated externally with some pro-
fessed water-proofing material—as "Gay's solution,"
"Szerelmey's liquid," "petrifying silicate," and
many others, the manufacture of which—*specially
to cure damp walls* — finds employment for no
inconsiderable number of workmen. Correspond-
ents of technical journals are constantly seeking
advice for the same evil, and are recommended
gas tar, soap made into a solution with water,
a covering of cement or roofing slates, and other
specifics.

The universal love for bricks and mortar, with all
their shortcomings and failures, will not probably for
some time to come be supplanted to any very great
extent by the introduction of any new, or the re-
introduction of any old and abandoned material for
wall building, no matter how strong, weather-proof,

durable, or economical it may be. Of late years there have been many substitutes invented, as Ransome's silicious stone, terra-cotta, and Taylor's facing bricks, all which, though equal in appearance and far superior to common bricks in durability and impermeability, can never be employed but to a limited extent, owing in a great measure to their cost, especially in remote districts, where the carriage would form a heavy item. On the other hand, the scarcity of good bricks, and the fact that they are only to be had at an extravagant cost—if quality is a consideration—the high price of labour, and the possibility that any building may experience a rest during construction, should a terra-cotta keystone or ornament happen to be fixed by any one but a recognised bricklayer, must eventually create a wide-spread feeling that the really most important portion of buildings—the walls—must in many instances be formed of something besides bricks, and the workmanship be capable of execution by others besides bricklayers. A material which should be as economical in any neighbourhood as bricks and mortar, should combine strength with durability, be proof against rains and frosts, withstand the injurious action of smoke, and be easy of application by unskilled labour, has, till within the last few years, been an acknowledged want, and that want, it is improbable, will be nearer realised for some time to come by anything but concrete. That Portland cement and lime concretes must and will, however, come into general use for building purposes eventually, does not admit of a doubt; at the same time local circumstances, with regard to natural productions, and the character and purposes of intended buildings, must still be the means in a large measure of deciding as to what shall be used in construction; and when the prejudice against concrete, like other prejudices, has died out, it will take its proper position as a valuable and, in most cases, an economical material—not to supersede everything else, but as a useful adjunct in building.

It is well worthy of notice, that of late the opposition to the use of concrete has resolved itself almost entirely into the question of "Æsthetics." The doubts about strength, durability, and other essential qualities, have given way to nearly the only remaining important difficulty, viz. how to treat it in such way that shall be reasonable and consistent, without imitating any other material, and without pretence of being something totally different from what it really is. Unfortunately for the reputation of Portland cement, it has clothed so many dilapidated and rotten structures, that the very name of stucco, with many, savours of a sham, and no other building material has probably been so well abused and less deserving it. For Portland cement has no rival, save such costly articles as granite and some other natural stones, for defying atmospheric changes, frost, rain, or the chemical action of smoke in large towns. There is nothing can be so easily and economically moulded into any shape or form for architectural purposes without distortion, and that will bear exposure to the vicissitudes of our own or any other climate; and there is no other material that can be applied to so many various requirements in building with such satisfactory results. The facing of concrete walls with Portland cement, only because it has been

found impracticable as yet to build them with sufficient
truth and regularity, and with even surfaces, so as to
require no further application afterwards of any finer
ingredients of the same nature, is in itself not so
much a sham as are the fronts of one half the houses
erected. Brickwork, more often than otherwise, is
stained to make it of an uniform colour, the joints
pointed with quite a different and superior mortar
to that used in construction, and the outside portion
of the walls built with bricks of a much better quality
than the inside. Terra-cotta in walls is scarcely ever
more than a few inches in thickness, nor would
any advantage accrue were it possible and usual
to have it the full wall thickness. Stone fronts
of many mansions and other costly buildings are
merely veneers of stone, backed with common bricks.
Where scagliola for wall-surface decorations, columns,
and pilasters; enamelled slate for chimney-pieces
and other fittings—both are openly recognised as
imitations of various marbles; parquetry made to
represent costly oak wainscoting, (the oak being the
thickness of a wafer!) deal is stained to represent oak;
fittings and furniture are made of fir and veneered
with the most costly woods; plaster and other ornaments are clothed in gilt and lacquer—where all these
and many more shams abound, with many, concrete would not be admitted, if, for appearance sake,
it had to be covered with a finer quality of the
same kind of materials of which it was composed,
and of which it would form an integral part. Possibly
there will be nothing invented or introduced as a
finish for concrete walls externally, that will require
eventually less maintenance, cost less in execution,

and defy climatic changes so well, as a coating or covering of Portland cement and sand, and in favour of this method the remark made by Mr. Reed, M.P., relative to the ugliness of iron-clad ships will well apply—" The beauty of a thing consists of its complete adaptation for the ultimate purpose in view;" and surely the ultimate purpose of a house is to give its inmates the healthy shelter necessary for long life and its enjoyments.

But although the irrepressible cement stucco may be, for the reasons stated, a suitable way of finishing concrete buildings, it does not follow tha no other method is open, the fact being that there is nothing so susceptible of different methods of treatment, or that affords an ingenious mind so wide a field for artistic effect and constructional development, as concrete. It is a significant fact that, although there have been, comparatively speaking, no end of inventions for making brick buildings damp-proof, nothing has as yet been offered in any shape for the purpose of rendering the same service to concrete walls (evidence in itself that the material requires no such extraneous aid), or as a remedy for incipient decay, a disease which attacks not a few stone buildings. The adoption of concrete so sparingly for constructional purposes, although it has been before the public for some years, may be however accounted for in various ways, besides the hereditary fondness for bricks and mortar, and the apathy displayed by so many as to the nature of the wall materials of houses, from a sanitary point of view; vested interest in brick-making, lime-burning, or bricklaying, and the policy of investing in concrete building appliances as

part and parcel of the ordinary plant of a builder,
have all had their share in the general result. Not
the least, however, of the causes, has been the exaggerated statements put forth by patentees, whose
sole interest has been to sell concrete machinery and
create royalties, and which statements have caused
disappointment to those who, without any previous
knowledge of the subject (or perhaps of building in
any form whatever), have been led to believe that the
use of concrete will effect a saving of 50 per cent.
in the cost of construction, and who have failed to
foresee that the materials of which the walls are built
can have but little influence on the value of the
carpentry, the painting, the slating, &c. As the cost
of the walls would be in an ordinary way perhaps
one-third of that of the whole building, the saving of
one-half their cost, which it is fair to assume might
under special circumstances be the case, would entail
a diminution of one-sixth only of the total outlay.
In another way, giving the cost of a cubic yard of
concrete, which it would seem was apparently a
matter not easy to get far wrong in, when the market
value of the materials are stated, and the proportions in which they should be used, the writer of the
statement and a well-known concrete builder made
it appear that the actual outlay would be 25 per cent.
below what it should have been, through calculating
sixteen bushels to a cubic yard ; and, most remarkable
of all, this statement not only remained undisputed,
but was copied into a well known builders' price
book.* It is to be hoped, in the builders' interest,

---

* Atchleys, 1868.

no very heavy contracts for concrete buildings were undertaken from estimates framed on this hypothesis. And the exaggerated statements as to strength and capabilities have only tended to make the uninitiated portion of the public disbelieve anything and everything connected with it, for concrete has been proclaimed a "salve for every sore," a "panacea for every evil," where a building was concerned. It has been stated over and over again to be ten times stronger than brickwork, which taken in a certain sense might be true; but, as read by anyone unacquainted with the subject, would be understood to mean that a concrete wall six inches thick was equal in strength to a brick wall five feet thick,— a comparison that has no semblance of reason in it. And, again, why concrete has found but little favour with many, is because it is assumed to be an untried material which age may deteriorate; but this is an assumption that will not bear reflection, for, whether formed with lime or cement, concrete is really a random rubble wall of cement or lime-mortar as the case may be, and in the construction of which the materials are blended together, and interlocked one with another, in a way that could not possibly be attained by the ordinary method of forming that class of work, and with the additional advantage that the cementitious ingredients must for obvious reasons be of the best and purest description, not composed of street scrapings, or garden soil, mixed with a modicum of lime, and with which a large proportion of the walls of brick buildings forming suburban London and many other places are built (and plastered as well—fortunately the ac-

tion of the lime destroys in a measure the fœtid
matter contained therein). Let an unprincipled
builder be ever so crafty, he cannot well build the
inside of his concrete walls of a different material
to the outside—cannot so easily face them with something fairly substantial, and leave the back a mass
of rottenness—a well-known and every day occurrence
with brick, rubble, or flint walls, where efficient supervision cannot be practised, or where houses are built
to sell for the highest sum they will realise.

With regard to being a modern invention, history
repeats itself in this as in most other things, for the
Greeks and Romans undoubtedly used it, and it exists
now in the foundations of many of their ancient buildings.

Palladio says: "Coffer work was made by taking
boards laid edgeways, according to the thickness of
the walls, filling the void or space between them with
cement and all sorts of small stones mingled together,
and continued after this manner from course to course;"
and Alberti also writes: "I have observed that
in other places the ancients, who were wonderfully
expert in making great works, followed different
methods in filling up the foundations. In the sepul-
chre of Saint Antonini they filled them up with little
pieces of very hard stones, each not bigger than a
handful, over which they perfectly drowned the mortar.
I have known other instances where the ancients have
much the same sorts of foundations and *structures* too
—of coarse gravel and common stones that they have
picked by chance, and which lasted many ages."
Here there is a positive assertion that concrete, not
only for foundations but for the walls themselves, must

have been known and practised many hundreds of years since.

Semple, in constructing the bridge over the river Liffey in 1753, used concrete in the foundations, and writes thus about it : " There are three different methods for making use of lime in such a work as this ; one is, to mix the roach lime (made liquid) with its proportion of sand and small stones in such a manner as may clothe every stone and particle of sand with it ; the second, to slack and turn them all up together like mortar ; the third, to lay each of the three, as it were, in thin layers, still observing the same proportion. Take which of these methods you please, provided the roach lime be, however, carefully and judiciously mixed with the stones and sand ; for if these materials are not equally mixed, how can you expect them to petrify and unite unto one solid mass ? But if they are properly mixed together the whole stuffing of this coffer will actually petrify, and become one solid compact substance, as hard and as closely united together as if the same was in one block or rock, and it will be many hundreds of years before the coffer (being in the water) will be in the least decayed."

The concrete walls of the fortifications of Badajos, in Spain, show at the present time the impression of the frames or boards used in their construction.

In this country we find that Corfe Castle is built of concrete. Reading Abbey walls were apparently faced with squared stones on the external sides, and the core, or backing, made of concrete, with the result that the stone appears to have perished, but the concrete, which was to have played a secondary part only, still remains. The existing portion of Old Sarum are some concrete

blocks or fragments of walls. Some parts of Roches-
ter Castle are concrete; and many walls of churches
and other buildings in the Eastern Counties appear to
have been constructed of a similar material.

In a work called "Santa Domingo and Haiti," by
Samuel Hazard, published in 1872, it is stated, with
reference to the buildings of Domingo: "The method
of making the walls is simple and economical; the
glutinous earth of the vicinity is taken and mixed with
lime, and sometimes, as in Cuba, with powdered stones;
frames of plank are then made in the desired form,
and these are filled with layers of the composition,
sand and lime being added. The whole is then
moistened with water, well pounded and kneaded, and
allowed to dry, when the mould being withdrawn,
leaves a firm solid wall, which, on exposure to the air,
becomes as hard as stone. Even the walls of the city
are built in this way, and the walls *of the older houses*
are constructed of a similar material."

Wherever we turn, therefore, we find that concrete can lay claim to having been a building material
long before many other artificial substances, now
recognised as legitimate and well-known materials of
construction, and the most earnest disciple of "high
art" and mediæval construction, can therefore rest
contented as to its antiquity.

As, however, Portland cement was unknown sixty
years since, all the concretes named had lime for a
cementitious constituent; but the excellent quality of
the same, the regularity and evenness with which all
the materials were incorporated, and the apparent
knowledge applied in the selection of the component
parts, resulted in the construction of buildings equal

to, or superior in strength and durability to, any of modern erection.

The idea of the cob or mud walls, common in Devonshire, Wiltshire, and other counties in the west of England, may have found an origin in the more pretentious early concrete buildings, as the mode of construction must have been very similar, viz. by forming moulds with rough boards the shape and size of the intended walls, and casting the mixture therein, in a soft mass.   The erection of these cob or mud walls is a common occurrence at the present time, and it is only just to say they are both drier and warmer than brick or stone erections; but the nature of the materials (mud, chalk, and straw), renders it necessary that both top and bottom sectional areas of such walls should be thoroughly protected from damp, and when this is done their durability is something marvellous, for buildings so constructed are often met with which have been in existence upwards of a century.   The residence of the late Sir William Fothergill Cooke, near Stockbridge, in Hampshire, is, perhaps, as extensive and complete a specimen of cob walling as can be found anywhere.

Concrete seems, however, to have lost prestige, or its merits to have remained unobserved, in this country for many generations.   This may have been in a great measure owing to the difficulty of obtaining quick-setting, properly prepared and power fullimes, and the absence of any appliances to facilitate construction, and to reduce the materials employed to a proper size when necessary.   How the builders of olden times managed these things, must be placed in the same category as many other wonders of construction—Stonehenge, for example—of which we are not likely to learn the secret.

The advantages possessed by concrete for constructional purposes appear to have developed themselves in modern times in the form of artificial foundations for sustaining heavy superstructures, in preference to the piling and planking, once generally adopted, but now universally abandoned in favour of the former method. No building of importance is at the present time erected (except the character of the subsoil is beyond suspicion and perfectly unyielding) without a certain amount of lime or cement concrete being employed, on which to commence building the walls; and the higher and more massive the building, the greater the thickness and width of concrete necessary.

And for the construction of breakwaters and similar works requiring large solid masses—larger than could be conveniently obtained and transported from the quarries if of stone—concrete began to be employed in the early part of the present century, but experienced the same opposition then, for that description of work, as it does now for house-building purposes.

In evidence given before a commission appointed by the Government, to inquire as to the best mode of constructing a harbour of refuge in the Channel, Mr. George Rennie, the engineer, said he objected to the use of any artificial or *inferior* materials, such as *concrete*, in any part of the breakwater. The invention of Portland cement, however, no doubt had a great deal to do with the ultimate adoption of concrete for works of this character, because it hardens quicker, and has far greater adhesive properties than lime, is unaffected by seawater, is eminently hydraulic, and does not deteriorate from age, thereby allowing large quantities to be stored ready for use.

Immense quantities of concrete have been used in the construction of breakwaters at Dover, Cherbourg, Alderney, Aberdeen, and many other places. In some cases it has been made into blocks from 50 to 350 tons weight each, then allowed to harden, and afterwards removed and lowered into their proper places, by means of lighters purposely constructed; and in other cases it has been deposited in a soft state, *in-situ,* or a combination of the two processes, has been adopted, blocks being used solely below high-water line, with block facing, and concrete *en masse* employed as a backing, above that level. In 1872 the British Fisheries Society's Report states, with regard to the Wick breakwater, that "it has been exposed to the very severe south-easterly gales and sustained damage, but the concrete blocks at the east-end of the works had not been injured, and this appears to show that solid cemented masses will resist the force of the waves better than the best masonry, however large the stones, and however strongly they may be bound together."

For a large amount of river walling at Chatham and Woolwich, concrete was extensively used at the beginning of the present century, but owing to some defect in the material, or negligence in the manufacture, it proved a failure, and probably led to its disuse for some considerable time; but as an instance of a quite different result, but under more adverse circumstances, we have the concrete sea-wall at Brighton, which has answered its purpose so well, after sufficient time has elapsed to thoroughly test it, that more is about to be built, as the severe storms of 1876-7, although seriously damaging the stone, had but little effect on the concrete walling.

We do not appear to possess any information of concrete being employed in this country for house building, in modern times, further back than about sixty years since, when Captain Moorsome, an engineer, constructed a house of lime concrete, about fifteen miles from Gloucester, and which is now, or was a few years since, still standing ; after that Mr. Ranger erected the College of Surgeons, and several houses and shops with a similar material, and with but partial success.   In 1859, M. Coignet took out a patent for a species of concrete, or beton, the name by which it is still known in France, but where the process of mixing the ingredients is somewhat different to that practised in this country, the lime or cement having first the sand and water added to it, so as to form a weak mortar, or "grout," in which state it is poured upon the gravel.

M. Coignet constructed a number of buildings on his principle, a full account of the erection of one, a railway station, appeared in the *Builder* and in the *Engineer* for the year 1859.   The material, called by the inventor "beton agglomérés," had ordinary lime as a basis, and the principal points observed were to avoid excess of lime, and to employ a very small quantity of water for mixing, then to subject the mass to a large amount of compression, or beating, which had the effect of rendering it dense, compact, and impermeable to water.

M. Coignet also constructed a quantity of sewers and several railway bridges on his method.

As in most other inventions, the claim to priority was disputed, and in this instance by a Mr. Buckwell, who constructed the basins in the nave of the Crystal Palace, and executed different descriptions of work in a very similar manner, previous to the date of M. Coignet's patent.

"Granite breccia," the name given to a material treated in much the same way, was extensively used in London, for paving streets, &c. a few years ago; and Victoria stone, which consists of Portland cement and fragments of granite cast in wood or iron moulds, and subsequently hardened by immersion in a silica bath, is largely employed as paving in the streets of London, and other places, with uniform success.

The Land Improvement Commissioners, in 1869, refused to grant the loan of a considerable sum of money for the erection of buildings in concrete on a large estate in the west of England; the matter was eventually referred to Mr. G. Godwin, the Editor of the *Builder*, who reported favourably on the subject, the result being the money was advanced for the purpose, since which time no further objections have been raised, but the Commissioners have to a certain extent recommended its use.

The *Builder, Field, Building News,* &c. have for many years strongly advocated the use of concrete in buildings, especially in homes for the labouring classes, and the latter journal not long since offered and awarded prizes for the best designs and treatment for a concrete villa; and so late as May 27th, 1876, speaking of the lingering affection displayed for damp and spongy brick buildings—for the majority are so, neither more nor less—and the prejudice

and antipathy existing against concrete, the *Builder*
says : " The common-sense technical view—that is,
the view taken by ordinary architects and builders—
of the application of concrete to the external superstructure of houses, is not based upon doubts as to
possibility or even economy. At the bottom of most
consciences lingers a half-regret—would that it were
impossible! There exists, and always has existed, a
love for buildings constructed of stone, and of combinations of brick and stone. There is an undoubted
prejudice against the look and even the feel of Portland cement. If indeed the realisation of some
sanitary Utopia depended only upon the general
acceptation of concrete fronts, the growing artistic
feeling of the people would experience a sensation of
revolt. . . The contemplated scarcity of bricks may
give an impetus to its extended use and inevitable
development; and a strike similar to that of the
Paris carpenters in 1840, which resulted in the effec-
tual introduction of iron floors, may be advantageously repeated in this country, in order to hasten
the exclusion, not only of naked timbers, but also of
naked iron, from every public and private building.
. . . There can be no doubt that the economical
application of ordinary concrete, such as ordinary
workmen can prepare, to the internal distribution of
ordinary buildings, would afford a great benefit to the
public generally."
A favourable instance of the application of concrete
to bridge construction was made in 1868 by the
Metropolitan Railway Company, near Gloucester-road,
who built one across a deep cutting of 75 feet span.
No iron was embedded in the mass, nor employed as

an auxiliary in any form. The rise in the middle—
the bridge being a single arch—was 7 ft. 6 in., and the
thickness at the crown 3 ft. 6 in. A load of 170 tons
was distributed over the whole surface, and a train
of loaded trucks, weighing 50 tons, was afterwards
drawn over it, and produced no sensible deflection;
when one side of the bridge was loaded, producing a
strain of $10\frac{3}{4}$ tons per superficial foot, the rise on the
opposite side was one-sixteenth of an inch; the width
of the bridge was 12 feet. It has since been removed,
for further railway extension.

A lighthouse was erected on the Isle of Jersey, in
1874, of concrete; in this case the concrete was
moulded into blocks, and the latter built up as
ordinary masonry; it is 135 feet high, and the rock
on which it stands is 109 feet above the level of the
sea. It was designed and erected by Sir J. Coode,
and is considered a successful application of concrete.

The invention of Portland cement, by Mr. Aspdin,
introduced a material eminently fitted for a constituent
in concrete; for although hydraulic lime, and the
more modern introduction of Roman, Medina, and
other cements, properly prepared, had been found
under certain conditions, and in the hands of those
competent to deal with them, capable of forming good
and substantial concrete walls, yet they had in this
country only received a limited amount of use, except
for foundations of buildings, backing of retaining-
walls, dock-walls, and similar objects; and although
the superior advantages possessed by Portland cement
should have quickly tended to stimulate concrete
building, it does not appear to have done so. But
there were many reasons to account for this. The

absence of any proper appliances to mould the form
required, the natural love for "honest bricks and
mortar," vested interests in brick-making, lime-burning, and all the accessories to the aforesaid honest
style of building—the prejudice always existing towards
anything bearing even the taint of newness or novelty
—simplicity of construction, and consequent doubts
as to efficiency—all helped to check the use of concrete
for a considerable time. One of the greatest disadvantages was the costly and clumsy method of forming
the frame or mould, which consisted of fixing temporary wood posts at all angles, and at intermediate positions on both sides of intended walls, and
nailing boards to the inner sides or faces of these
posts, to form a matrix or mould of the shape and
size required for the intended wall. This process
necessitated no inconsiderable amount of both labour
and materials for temporary purposes alone, and
added so much to the ultimate cost of building that,
except under very special circumstances, it virtually
prohibited the employment of concrete for house-con-
struction ; but in 1865, the notion of using movable
frames or panels, so that when one portion of the
walling enclosed within those panels had been formed,
the latter might be removed and re-fixed immediately
above the parts already completed, and so on *ad
infinitum* to any reasonable height, occurred to Mr.
Joseph Tall, who patented the first, and for some
time the only, form of portable aid to concrete building
known. As, however, concrete came into more
general use, and its applicability to various purposes
better understood, so other appliances were invented
and patented, but all, or nearly all, modifications of

the same principle first adopted by Mr. Tall, viz. of employing fixed standards or guides to regulate the thickness of the walls to be built, and also determine their correctness and precision, and of movable panels which are temporarily attached to the former, and form the matrix for the intended walls.

The usual amount of opposition experienced by all who introduce anything novel or ingenious soon made itself·apparent in the endeavour of the Metropolitan Board of Works, who claimed the right to apply the rules and regulations in force within the Metropolitan district for the construction of brick and stone erections, to the construction also of concrete buildings, the nature of the materials in the latter, and the method of using them, being perfectly different to ordinary erections. The principal point at issue appeared to be not the most vital part, viz. the fitness and correct proportions of the ingredients, but whether " bonding " in the ordinary sense of the word, as applied to brick and stone work, was to be enforced in concrete work.

The fact was not disputed that, as concrete walls have no one part weaker than another, it was impossible to state where or how " bond " should be employed; but as no provision had been made nor contingency provided for other than brick or stone erections, it was maintained that the rules laid down for the latter should be observed in the former mode of construction. Eventually the matter was com promised by the Board of Works granting licenses, under certain conditions, to builders to use concrete.

CHAPTER I.

DIFFERENT MODES OF EMPLOYING CONCRETE, AND
SYSTEMS OF CONSTRUCTION.

CONCRETE, as now used, is simply a mixture of either
Portland cement, ground lime, Roman cement, or
other cementitious material, with gravel, river ballast,
stone chippings, burnt clay, furnace cinders, slag from
iron ore, crushed flints, shingle, broken bricks, tiles,
stone or débris from old buildings, or some similar
materials to which water is added ; and when the mass
becomes thoroughly incorporated, placed in a semi-
fluid condition in moulds constructed with common
boards, or with patent appliances manufactured for
the purpose. The main constituent of the concrete is
called the " aggregate," and the cementitious ingredient is improperly designated the " matrix."

It will be seen that on the proper adjustment of the
moulds depends entirely the correct shape and development of chimneys, walls, and other structural portions of the bnilding, and that the character of the
work depends, therefore, entirely on two things—the
precision and correctness with which the mould is
formed, and the quality and relative proportions of
the materials and their proper manipulation. There

is, however, another method of building with concrete, viz. by casting it into small blocks, allowing these to become hard by exposure to the atmosphere, and using them in a similar manner to bricks or blocks of squared stone.   This plan, called the " block " system, to distinguish it from the one more generally adopted, and designated the "monolithic," or continuous system, has some advantages, the principal of which is that Portland cement concrete increases in strength the more it is pressed or rammed, and that by using strong wood or iron moulds a large amount of pressure can be obtained; whereas the construction of walls, as first described, will not allow of this, except the moulds or frames are made exceedingly strong, but which would at the same time render them too heavy and cumbersome to be portable.   Mr. Reid, in his treatise on " Concrete," in enumerating the advantages of the " block " system, says :—

" The labour required to fill, ram, and empty the moulds can be of the cheapest kind, and even women and boys may be employed for that purpose when necessary.   The blocks could be made at convenient intervals, and allowed to remain for months or years before being used.   Less water can be used, and a larger proportion of gravel or other aggregate introduced into the block.   In districts where employment of labour is intermittent and uncertain, the unemployed might be advantageously engaged in moulding blocks, which would readily find a market for house-building and other purposes.   No imperfect block could be by this arrangement used, or a bad quality of cement employed, for sufficient time would be permitted to detect the shortcomings of the one and the other."

These acknowledged advantages are in practice, however, more than balanced by the following disadvantages :—

1st. Buildings constructed with concrete blocks necessarily require skilled and therefore expensive labour, equally the same as if the blocks were natural stone, or clay bricks; and although it has been attempted to prove otherwise, it is only reasonable to assume that blocks artificially made of concrete cannot be formed into walls in a proper manner with any less judgment or skill, because they (the blocks) are made in a different way, and with different materials, than ordinary bricks are.

2nd. When concrete blocks are made from ballast, gravel, flints, or other hard materials, they cannot be sawn smaller, as stone could be, or "cropped," as bricks can; and where cross-walls intervene, chimneys project, or other causes necessitate bond-stones and short closers, or where indents, toothings, corbels, setoffs, and other irregularities common to all buildings occur, blocks of the necessary size and shape would have to be moulded.

3rd. The strength of the monolithic system is not attained (unless cement be used for all joints), and the advantages, therefore, of increased density and resistance to crushing force, is counterbalanced by the introduction of weak mortar-joints.

4th. The cost of labour alone in laying the blocks would be more than that of making and depositing the concrete in place, on the monolithic system, and the great advantage of diminution of skilled labour required by the latter method is not gained by the former. Although concrete blocks might be faced with a thin coat of cement previous to use, and so obviate the necessity of cementing or otherwise facing the work after erection, yet the cost of treating small blocks in this way would be greater than cementing or stuccoing large finished surfaces of walls.

5th. The formation of chimney flues and throats, and the twisting and turning necessary in most cases must be obvious, is a matter of no small difficulty where the blocks cannot be cut or altered in shape or size.

The objection to the monolithic system, that bad materials may be used, is applicable to any kind of building, and which can be remedied equally as well in one case as in others, viz. by strict supervision, and testing the materials employed.

But several combinations of the block and the monolithic systems have been introduced. The Broomhall Company have for many years supplied a superior kind of facing brick, originally patented by Mr. John Taylor; the walls are faced on both sides with the bricks, and the concrete is deposited in the cavity between, the latter process being repeated every course. The advantages possessed by the patent over ordinary bricks are,—their light weight and peculiar section, which latter secures efficient bond with the concrete, and obviates the necessity of "headers" to secure that purpose, consequently enables much thinner walls to be constructed than otherwise could be. Walls of any thickness above six inches can be built by this system, the difference consisting merely in the amount of concrete core deposited between the inner and outer brick facings. Where these bricks are used, concrete building appliances are, of course, unnecessary, and ordinary bricklayers' plant and scaffolding only needed ; buildings constructed on this principle are substantial, durable, and weather-proof ; but the disadvantages are, the necessity of employing skilled labour in addition to concrete mixers, and an ultimate cost exceeding that of ordinary brickwork.

Sellars' patent blocks consist of coarse sand and Portland cement cast in moulds, the blocks so formed being subjected to great pressure to secure density, increased strength, and smooth surfaces ; the blocks are cast hollow, thus economizing both materials and cost of carriage ; and the hollows or cavities are, during the process of building, filled with rough stones or with concrete ; the ends of the blocks are fluted or waved, to admit of their fitting into each other, and a special kind of interlocking brick is also made for division or party walls.   The patentee uses the somewhat singular title of "tent-building" for work executed with these blocks where no cementitious material is employed for bedding the joints, and the cavities are not filled up ; and " rock building " where mortar or cement is used in the ordinary way, and concrete or rough stone as packing.   Taking the patentee's own description of small cost and other advantages realised by this process, we are told : " Houses of double the strength and twice the durability of common clay-brick tenements, with a great saving in the cost of construction of buildings, can be effected by using these fluted hollow blocks and interlocking bricks made of hardened concrete. Taking common clay brickwork in any country as the standard, a profit of £220 out of a £600 estimate can be realised if the blocks and bricks be made by hand, and

£403 out of a £600 estimate if made by steam or hydraulic power."

The details of cost to effect this are stated to be : " Hollow stone blocks built as hollow walls, cost 5s. per " super yard + 3d. cartage, + 3d. setting, equalling " 5s. 6d. per super yard for 9-inch work." 5s. 6d. per super yard for 9-inch work is equal to 22s. per cubic yard, or £12 7s. per standard rod; and practical builders must themselves solve the enigma of the immense saving, and also the possibility at the present time of building walls for 3d. per super yard, or 11s. 3d. per standard rod !

The same objections that exist against the block system generally are also applicable in this instance; and the difficulty that at the present time exists, and for many years to come will probably continue to exist, of inducing masons, stone-setters, or bricklayers, except in individual cases, to feel any interest in new materials and improved systems of construction, is one great disadvantage possessed by every form of block building known.

Lascelles' patent concrete slab-buildings, except that the slabs are made from Portland cement and sand or gravel, can scarcely be designated " concrete constructions," as the ordinary woodwork of half-timbered erections are necessary, and to which woodwork or quarters the concrete slabs are fixed on the external side by ordinary screws; the slabs having rebated horizontal joints to prevent moisture penetrating.

Here, again, it is difficult to discover how great economy can be practised when the timber framing of itself must form no inconsiderable item of cost, consuming valuable materials and necessitating skilled labour in erection ; but the patentee proposes to remedy this by using concrete studs or framing, concrete floor joists, and concrete rafters, and the most enthusiastic disciple of concrete construction can scarcely wish more. Apart from this, however, using timber framing for the walls, and wood rafters to carry similar slabs for roofs, as employed for the walls, it is scarcely reasonable to infer that the cost can compete with monolithic construction ; and the strength and durability of the buildings must be a matter depending wholly on the lasting properties of the skeleton, or wood framing. The entire principle is wrong, for the concrete is not considered of any importance in a structural point of view, which is really its greatest feature, but is entirely subservient to an inferior material. For the finishing internally, the usual process of lath-and-plaster, or other known methods, are to be adopted. If we are to take the patentee's assertions as contained in his pamphlet, all other kinds of construction, whether concrete, brick, or stone, must eventually succumb to his method, as we are informed that " buildings on this principle re-" quire no scaffolding—no excavation—no slates or tiles " —no metal coverings—no concrete filling in—no sunk " foundations—no door frames—no window frames— " they are fire-proof, weather-proof, and vermin-" proof—are non-conductors of heat and cold—are " indestructible—can be built in winter—occupied as " soon as built—erected very quickly and removed, con-" veyed any distance, and re-erected." Assuming that the qualifications concrete slab buildings are described to possess are fairly stated, the ordinary brickmaker,

lime-burner, and concrete builder, will, with Othello,
find his occupation gone, for they leave but little to be
desired. Wilkinson's patent concrete stable paving,
Dennett's fire-proof construction, and other patented
forms of concrete as applied principally to the formation of arches, vaults, floors, and pavings, consist of
selected materials as aggregates, specially adapted for
their intended purposes, and various cementitious
constituents applied in such proportions and in certain forms as tend to give the best results. But whatever the description of concrete building, whether
block, monolithic, or a combination of the two, the
proper selection of materials, and their proper manipulation are of the greatest importance. Many
failures in concrete work have occurred through in-
judicious treatment, and too much faith being placed
in what is often sold for Portland cement ; and because
good cement possesses immense powers of cohesion
and adhesion, many people assume that no matter
what the aggregate may consist of, or how unfitting
for the purpose, the effect will be equally as good as
when the greatest care is taken in selection ; and in
many other instances, provided the name of Portland
cement is given to a compound more or less composed
of ground lias lime, slag dust, &c. it is taken for
granted to be all that is desired. In the thorough
amalgamation of the ingredients, also, often no proper
supervision is exercised, or the operation is entrusted
to incompetent workmen, whose delight is to find a
cement that rapidly sets, and who probably, after
mixing, expose the concrete an unreasonable time
previous to being deposited in place till consolidation
has fairly commenced, when more water is added, and

the universal treatment applied to lime mortar under similar circumstances, is assumed to be the correct way to deal with cement concrete. Occasionally the materials are incorporated just as little or as much as the usual class of workmen employed thereat can arrange to perform in proportion to the amount of supervision they experience. We are told that as hanging scaffolds are often suspended to the walls of monolithic concrete buildings during construction, and that, as on the stability of the latter depend the lives of the workmen, they—the workmen—will not fail for their own safety to prepare the concrete in a thorough manner ; but, as everyone engaged in building operations know full well, but little weight can be attached to this argument. When sheer negligence and ordinary want of even common care is the rule, it is not to be wondered at that in buildings which have been erected some little time, cracks and fissures are discovered, rains penetrate, other evils prevail, and concrete is declared to be a delusion and a snare.

CHAPTER II.

AGGREGATES.

FIRST, then, in selecting the aggregate, and whether for block building or for monolithic building, the same quality of materials and careful workmanship are necessary; but assuming that the last-named method is decided upon, the description of aggregate must depend largely upon local circumstances, as haulage forms an important item in the cost of every kind of wall building. In the neighbourhood of iron mines, slag, the refuse from iron ore, can be readily procured, often for nothing, or, at most, at a nominal cost, and probably no more fitting material exists than this for making good sound concrete, and, if the finer or sandy parts are in too great a proportion, then their abstraction give an ingredient superior to many natural sands for use in hair mortar, or common mortar for ordinary purposes, or for mixing with cement for external stuccoing. The colour imparted to stucco is not, however, everything that could be desired, but as a rule it is uniform, and not so objectionable a tint as many pit sands create. In the neighbourhood of quarries, stone chippings of any description are useful as an aggregate, the only disadvantage in their use being that in shape they are often flaky, and do not therefore fit into each other to form a mass dense enough for sound work ; in fact, the best shape for all portions of the aggregate is that possessing the greatest number of angles, and their respective surfaces but little superficial area. Flints from chalk strata, or collected from the surface of the land, in places where they are abundant, also make a good aggregate, but breaking them with a hand hammer such as stone-breakers use, does not create sufficient portions of irregular size, and this remark applies to every other kind of material, for the most irregular size, but not exceeding what would pass through a ring two and a half inches in diameter for the thickest walls, but less for thinner walls, and not finer than coarse Thames sand, produces the best concrete. Pea gravel, or gravel of the uniform size of a pea, has been used for concrete with indifferent results, on account of the absence of any smaller particles to fill the interstices between the larger, and the much greater amount of superficial area requiring to be cemented together, than if some portion were of large dimensions. The only way of producing materials fitting for concrete from flints, boulders, old stones, brickbats, &c., and that require no further preparation, is by crushing them with machines made for that purpose, none of which are superior to Blake's stone-crusher. These machines are made from two-horse power upward ; the smaller sizes can be driven by horse gear, but for cost of production and general utility, those capable of being worked by a four-horse power steam-engine, and costing about £150, are the best, and will crush from 20 to 40 tons per day requiring four men to supply, feed, and remove the materials. Where a crusher can be employed with advantage for other purposes as well, such as breaking stones for repairing or forming roads, or stone for lime-burning, it will quickly repay its cost, as the working parts are easily renewed at a small cost when worn out. Where flints are used for concrete—chalk, clay, or argillaceous matter of any description must be removed by washing, and it is better to wash them previous to crushing, as not only do stone-breakers act more efficiently and productively where the materials are clean, but in washing after crushing, the sandy portions are liable to be lost, or washed away; but assuming this latter plan is adopted (and it may from various causes often occur), it is best to wash the crushed aggregate on a wood platform, or on loose boards laid on the ground with a slight inclination, to allow the water to run away, but without sufficient force to carry with it the sandy and finely-crushed portions. The "mixing-board" (see page 76) on which the concrete is intended to be mixed can be used for this purpose, and if performed at the same time that the concrete is required, will save unnecessary removal, and economize labour; where the work is extensive and sufficient mixing-boards are provided, one set of workmen can be employed washing the aggregate and another making the concrete.

Thames gravel—or ballast, as it is more commonly called—is one of the safest materials that can be employed for concrete, because its particles come to hand ready washed from all impurities and usually of a very irregular and suitable size; the finer portion is a sharp coarse sand, especially adapted for mixing with cement to form stucco (should it exist in too great a proportion for concrete purposes). Where the cost of haulage is not excessive, Thames ballast is a profitable aggregate, as there is no labour in washing, and it can often be obtained fitted for mixing with the matrix without any previous preparation.

River gravel makes a more or less fitting aggregate according to the velocity of the streams in which it may be found, and which is essential to free it from all soluble matter, but a great deal of river gravel is overcharged with sand of much too fine a character. The quicker the current, as a rule, the better and cleaner the gravel obtained from the river beds, and consequently the more suitable for concrete; but the banks and meadows adjoining many rivers contain, however, more fitting materials than the rivers themselves, and any impurities can be readily removed by well washing the gravel on the spot where water is abundant.

Brickbats, clinkers or fused bricks, broken pottery, tiles, drain pipes, terra-cotta, and other débris from brick yards, when crushed, constitute a favourable aggregate, and the same remark applies to the walls of old buildings, whether of stone, brick, or flint; and old roofing tiles, paving stones, and paving bricks, assist equally well for the purpose, provided moss, lichen, &c., are removed; but slates are, when broken up, too thin and splintered in form to be of advantage, and should be avoided; the mortar in old walls, more especially plastering mortar, should be taken out, or at least the greater portion of it, especially if it is of inferior character, or the matrix is to be Portland cement. For a concrete aggregate

the materials in old buildings are often most profitable, as they are fitted for but little else; and in
cases of rebuilding on same sites, the saving in
haulage alone is no inconsiderable proportion of
the value of reconstruction. In many instances,
where no convenient place exists whereon to deposit the débris from old buildings, it has to be
carted long distances, and at most is only service-
able as a foundation for new roads, or for sub-
draining garden paths or carriage drives. Here,
again, to make sound concrete, "Blake's" or some
other stone-breaker is necessary for preparing the
materials, and if building sand is scarce a double
purpose can be served, for the old bricks can be
crushed first, and then passed through edge rollers
or a sand-grinding machine, and sand suitable for
plastering purposes can be obtained : a steam-engine
will perform both operations of crushing concrete and
grinding sand at the same time, and economize the
cost of production.

    Pit gravel is admissible as an aggregate, but it is
in most cases found in combination with clayey or
argillaceous matter, and where this occurs it is necessary to thoroughly cleanse it by washing; this will
often abstract the finer portions to some extent, and
it may be found necessary to add coarse sand or
shingle, to provide a sufficient portion of smaller
size. It may be taken as a safe guide to the fitness
of pit gravel for concrete, that the more binding
or the better adapted for footpaths or carriage
roads, the less fitting is it in its natural state for an
aggregate.

    Sea gravel or beach will make a good concrete, for

although its edges are rounded by attrition and therefore unable to interlock into each other as more
angular materials would do, it has the advantages
generally of being irregular in size, perfectly free from
any impurities, and containing a suitable proportion
of coarse sand; and as Portland cement sets equally
as hard mixed with salt as with rain or spring water,
it has no disadvantage in this respect, and we therefore naturally find concrete, from the facility with
which the aggregate can be obtained, a good deal
in use at seaside places, numbers of large houses
having been erected at Folkestone, Clacton-upon-Sea,
Brighton, and other watering-places. It is asserted
that the salt contained in the sea beach in no way
tends to create dampness in concrete walls, nor do
houses built with it appear to be affected in that way.
For sea walls, concrete in a monolithic form has been
found—as before stated—to answer more effectually
than stone, and it is self-evident the cost must be
only a fractional proportion of the latter material.

Clay, if thoroughly and properly burnt, is a very
good aggregate, but does not find favour so readily as
many other materials, and for various reasons : the
cost of burning,—the irregular character of the clay,
unless great judgment is exercised in obtaining, tem-
pering and burning,—the probable excess of either
finer portions, or the reverse, thereby necessitating
screening or crushing—and the facility in almost any
neighbourhood of obtaining either one or other of
the materials previously described, tend to render the
operation of digging and burning clay unnecessary.

Mill cinders from factories, coke from gas-works,
and almost any mineral material that has been ex-

posed to great heat and afterwards retains the shape
and hardness necessary for good concrete, is an excellent aggregate, and for all of which Portland
cement has a strong affinity.

Chalk, if properly selected, will answer the purpose
also sufficiently well for many descriptions of work,
such as boundary walls, sheds, &c., but the soft upper
bed can scarcely be recommended as suitable. Mr.
Reid, in his work on "Concrete," says that as chalk
contains 12 per cent. of latent moisture, it should
first be deprived of this by submitting it to a temperature of 120 degrees of heat ; this is, however, not
always practicable, having regard to cost, and for
other reasons; and although it may not be so good
an aggregate as many others, yet chalk in its natural
state if left exposed to the atmosphere for a few
months in the summer season, will make concrete
walls of great strength and durability; in fact, Port-
and cement will indurate the chalk, and render it
proof against the most severe frosts.

These assertions have been sufficiently proved by
constructing a specimen of chalk walling 9 inches
thick, 4 feet high, and in form of a right angle on
plan, each side measuring 5 feet in length. A chain
attached to one side at a point half-way in height
from the ground, and half-way—in an horizontal
line—from the angle, was strained six months after
construction with a direct pulling force of 30 cwt.
without fracture, the entire weight of the materials
used in forming the wall being 28 cwt. This specimen of walling, without any coat of cement, or other
protection, was found to have defied all weathers six
years afterwards.

But the finer portions of the chalk should, however, be abstracted by screening it through a ¼-inch mesh sieve or screen, and adding instead thereof a sufficient quantity of clean coarse sand or shingle, or finely-crushed brick material. But although all the aggregates that have been named if properly treated will answer well to all intents and purposes, yet some regard should, if possible, be paid to the description of matrix it is intended to employ; Roman cement, for example, being quick setting, if used as a matrix with burnt clay or broken bricks as the aggregate, has its powers of setting hastened by the absorbing nature of the latter, and is therefore not so good for the purpose as a less absorbent material would be, and, on the other hand, the slower setting properties of lime causes a spongy aggregate to be favourable to hardening, and hastens what would sometimes be a tedious process. Portland cement, however, possessing the happy medium of setting steadily, much slower than Roman cement, and quicker than lime, can be applied with nearly equal advantage to any aggregate, although it is only reasonable to suppose that aggregates possessing a moderate amount of porosity, as slag from iron-ore, oolitic limestone chippings, or almost any kind of broken stone, give the matrix a power of adhesion superior to the smooth glass-like surface of flints.

Where the aggregate is gravel, sea-beach, or anything that comes to hand sufficiently small without breaking, no choice exists of adapting the size of same to the purpose for which it was intended, but where it has to be crushed or screened, the size should be regulated accordingly,—thus for 9-inch

walls and upwards no portion should be larger than
would pass through a 2½-inch ring, or screen; for
walls between 4 and 9 inches thick, and for arches of
floors and roofs, tanks, pits, steining of wells and
similar objects, not larger than 2 inches; and for
walls 4 inches in thickness and under, not more than
1½ inches: but this rule may be subject to digres-
sion, as a preponderance of the finer kind in the
aggregate would allow a portion of the remainder
to be larger than the dimensions given, without de-
tracting from the strength of the concrete, but rather
tend to improve it. Should the aggregate, however,
approach to an uniform size throughout and contain
but a limited amount of sand or shingle, then these
specified dimensions should not be exceeded, but somewhat reduced instead.

An important point in connection with the aggregate is to have it of one uniform consistency, not
at one time overcharged with sandy ingredients,
and at another almost entirely composed of coarser
portions; and the due proportion of both fine, coarse,
and intermediate sizes has a great deal to do with
both the ultimate strength and resistance to moisture of the concrete. If an excess of sand is used,
the concrete really becomes more a mass of coarse,
spongy mortar; and unless a larger quantity of
the matrix is employed, has none of the charac-
teristics of good concrete, but is pervious to damp
and incapable of sustaining any great strain or pressure; on the other hand, if an insufficient proportion
of sand is the rule—the larger material not having a
relative amount of smaller to fill the interstices and
form a connecting medium to the bulk of the aggre-

gate—the concrete becomes brittle, rotten, and porous, and the faces, or surfaces of walls so built have an honeycombed appearance; nor will any extra quantity of the matrix compensate for the absence of a properly-selected aggregate. It has been attempted to lay down certain rules for ascertaining the correct proportion of sandy particles in a given quantity of the aggregate, but practically it is for many reasons impossible to do this, and the knowledge can only be gained by practice and experiment. Probably the general rule is to employ an excess of sand or finer portions, as the results of so doing are smooth and even surfaces, and consequently an appearance of stability, not possessed by concrete walls in which the aggregate has been judiciously selected.

It is a common error to use an inferior aggregate for foundations of walls; the materials are often good enough in themselves, but with no care bestowed on their size, and in freeing them from deleterious substances, and although the ground on each side may be capable of averting any tendency of the concrete to bulge, yet it has more heavy pressure to sustain than any other portion, and therefore should be composed of equally as good materials; besides this, the foundations of buildings are exposed to dampness and rainfall, and should the surface of the ground at any future time be lowered, the weakest portion of the building may be exposed without any form of support. The fall of a large concrete engine-shed on the Metropolitan Railway a year or two since was proved to be owing to the foundations having been levelled for the superstructure in some places above the surface of the ground : the walls of the building were of cement, but the foundations were of ordinary lime concrete, and the latter, being unable to sustain the great weight put upon them, were crushed, and the walls they were meant to sustain fell to the ground.

In bestowing what may be sometimes considered unnecessary labour in cleansing the aggregate, and obtaining it of suitable proportions, where the work to be executed is of an unimportant character, it must be borne in mind that it is economy of cost to do so, because an efficient aggregate will make as good concrete with half the volume of the matrix that would be necessary with an aggregate of an unsuitable nature ; and the value of the cementitious material thus saved is more than the extra cost incurred in procuring proper materials ; moreover, concrete with a clean aggregate improves by age, but with a bad one it deteriorates. A good from a bad matrix cannot be always detected, even by experienced persons, without the aid of special machines made for that purpose ; but with aggregates it is different, the tact and discrimination necessary for forming a correct judgment as to their value for concrete purposes can be gained with a moderate amount of application and experiment, and when this is the result, the first step towards a knowledge of concrete construction is the result.

CHAPTER III.

MR. REID appears to have been the first, in his work on "Concrete," to have used the word "matrix" for the cementitious ingredient, although it is somewhat difficult to understand why this should be so, a matrix generally being supposed to mean "a mould or cavity," and therefore would apply more correctly to the space or void in which the concrete is to be deposited.

There are a considerable number of cements, and both ordinary and prepared limes, which can be employed as matrices, some used in conjunction with other materials, and incorporated when required for use, as lime and puzzolana, and lime and trass, called " compound matrices ; " others used simply as manufactured, as Portland and Roman cements, and called " simple matrices." These latter are, from ease and facility of application, now generally adopted ; and Portland cement, possessing so many advantages and few defects in comparison with others, has nearly driven all its competitors out of the field.

An objection to Roman cement is that it sets too rapidly for ordinary concrete purposes, although this is taken advantage of in special circumstances, such as in tidal or other hydraulic works, where the water can only be stayed a limited time, insufficient perhaps to prevent a slower-setting cement from being washed away. Other objections are that it loses strength if kept only a short time after being manufactured, and only a limited quantity at a time can, therefore, be retained for use; and it will not, moreover, bear the same proportionate amount of sand or aggregate that Portland cement is able to, without losing a much greater proportion of its powers of adhesion. Mr. Grant, in a paper read before the Institute of Civil Engineers, says that " Roman cement, although only two-thirds the cost of Portland, is about one-third of its strength, and is therefore double the cost measured by strength." However fitting, therefore, it may be for special and peculiar purposes, it cannot compete with the latter for concrete purposes in ordinary buildings. Roman cement was invented by Mr. Parker, and originally called " Parker's cement." The patents of protection are dated respectively 1791 and 1796, and the latter was probably for an improved method of manufacture; the name " Roman " is supposed to have been given it by its inventor, under the impression that he had discovered the method adopted by the Romans in the manufacture of their mortar; and so general in use and high in estimation had Parker's, or Roman, cement at one time attained, that during Sir Robert Peel's premiership it was proposed to tax foreigners dredging for the stone from which it is made. Mr. James Wyatt was one of the first to introduce it to public notice, and it was first sold by Charles Wyatt and Co., Bankside, London,

at 5s. 6d. per bushel. Medina cement is prepared in
a similar manner and from similar materials as
Roman cement, except that the former is produced
from stones found at the Isle of Wight, and the latter
mostly on the coasts of Kent and Essex; and the
same objections that apply to one for concrete purposes, apply also to the other. Portland cement was
invented by Mr. Aspdin, and patented October 21,
1824; it is supposed, even now by some, to be made
at Portland from the well-known Portland stone, but
it in reality owes its name to the fact that when of
good quality it is similar in colour and appearance if
trowelled to a smooth surface, as rubbed Portland
stone. This cement appears to have attracted but
little notice for some considerable time after its invention, and its now well-known properties to have remained undeveloped for many years. In a book
entitled the *Bricklayers' and Plasterers' Guide*, published in London in 1829, five years after its introduction, it is not even mentioned; and in the earlier
volumes of the *Builder*, published twenty years subsequent, it is scarcely noticed, and its proper position as
a building material barely recognised. The merits it
possessed in so many ways were, however, brought
prominently before the public by its use in the construction of the Thames Embankment and the main
drainage works of the metropolis, and in an elaborate
series of experiments made by Mr. John Grant, C.E.,
in 1859 and subsequent years, with the view of testing
its adaptability for work of so important a character.
The effect of this was to stimulate manufacturers to
increase the quality of their cement, and also enabled
architects and engineers to specify that it should be of

a certain definite strength, and by the invention of
machines to determine this, to ensure what had before
been only guesswork. The result was soon apparent
in the demand for cement of a much higher character
than had been previously known, and in the capability
of the manufacturers to satisfy this demand. The
original specification for the southern high level sewer,
constructed in 1859, stated that—"The whole of the
cement to be used in these works, and referred to in
the specification, is to be Portland cement of the very
best quality, ground extremely fine, weighing not less
than 110lbs. to the striked bushel, and capable of
maintaining a breaking weight of 400lbs. on an area
$1\frac{1}{2}$in. by $1\frac{1}{2}$in., equal to $2\frac{1}{4}$ square inches, seven days
after being made in an iron mould, and immersed six
of these days in water." The Board of Works' test in
1870 was specified thus—" The whole of the cement
shall be Portland cement of the very best quality,
ground extremely fine, weighing not less than 112lbs.
to the striked bushel, and capable of maintaining a
breaking weight of 350lbs. per square inch seven days
after being made in a mould and immersed in water
during the interval of seven days." It will be seen
that the later specification compelled a cement nearly
double in strength to that at first required.

The inventor of Portland cement describes himself
as " Joseph Aspdin, of Leeds, in the county of York,
bricklayer," and although he calls his invention
" Portland Cement," in the specification he alludes to
it more specially as " An improvement in the modes
of producing an artificial stone." The full text of the
specification, which could not well be more brief, is
as follows : " My method of making a cement or arti-

ficial stone for stuccoing buildings, waterworks, cisterns, or any other purpose to which it may be applicable, (and which I call Portland cement) is as follows : I take a specific quantity of limestone, such as that generally used for making or repairing roads, and I take it from the roads after it is reduced to a puddle or powder, but if I cannot procure a sufficient quantity of the above from the roads, I obtain the limestone itself, and I cause the puddle, or powder, or the limestone, as the case may be, to be calcined. I then take a specific quantity of argillaceous earth or clay, and mix them with water to a state approaching impalpability, either by manual labour or machinery. After this proceeding I put the above mixture into a slip pan for evaporation, either by the heat of the sun, or by submitting it to the action of fire or steam conveyed in flues or pipes under or near the pan, till the water is entirely evaporated. Then I break the said mixture into suitable lumps, and calcine them in a furnace similar to a lime kiln, till the carbonic acid is entirely expelled. The mixture so calcined is to be ground, beat, or rolled to a fine powder, and is then in a fit state for making cement or artificial stone. This powder is to be mixed with a sufficient quantity of water to bring it into the consistency of mortar, and thus applied to the purposes wanted."

The materials now employed for the manufacture of Portland cement are chalk and clay of a peculiar character. The alluvial clay obtained from the bed and banks of the Thames and Medway, and the narrow creeks leading therefrom, is a favourable one for the purpose; and as chalk to an almost unlimited extent can be procured near at hand, and good facilities for transit

exist by both rail and water, the manufacture of
cement has become one of the most important indus-
tries, especially on the Medway, thousands of tons
being manufactured yearly, and hundreds of workmen
employed in its production.   The proportion of chalk
to clay generally used is about 60 to 70 per cent. of
the former, but this depends, however, on the nature
of the chalk (less being used in the grey chalk districts
than in the white), and also upon the specific gravity
required.   The process consists of intimately mixing
the ingredients in large vessels by means of revolving
metal blades or knives till the whole is brought into a
creamy mass ; when this is effected it is run off into
tanks or holders, and the clear water allowed to settle,
when the latter is withdrawn.   The chalk and clay,
when sufficiently air-dried, is cut into lumps and removed to the kilns or ovens, where it is subjected to a
very high temperature.   When thoroughly burnt, the
lumps are allowed to cool, then ground to a powder,
and finally packed into casks or bags for use.

It is sometimes usual to fill the latter with the
cement as it leaves the grinding-mill, but this practice should be avoided, as no equality of strength can
be thereby ensured, and the extra cost of allowing it
to lay in bulk previous to packing is but trivial compared with the benefit obtained of having an uniform
strength.   The cement should be finely ground, as the
coarse grains, unreduced to an absolute powder,
although not possessing the same amount of tendency
to expand or blow as coarsely-ground lime, yet have
little or no cohesive or adhesive properties.   The
coarsely-ground cement weighs more than the finely-
ground, and this has held out an inducement to manu-

facturers not to grind it fine, as great reliance is placed upon, and specifications as a rule enforcing, a heavy cement. The Americans use a test-sieve having 80 meshes to the lineal inch, or 6,400 to a superficial inch. No doubt the greatest tensile strains have been obtained from cement of high specific gravity; but it is doubtful whether for concrete purposes in ordinary buildings, any advantage is gained by using it heavier than 112 lbs. per bushel; and under any circumstances it is better and safer not to employ it for at least seven days after delivery on the works, and even then it should be spread out on a wood floor, exposed to the air, but kept perfectly dry and occasionally moved with a rake or hook, to allow the air to penetrate and slake any coarse particles it may contain. Portland cement can be manufactured to weigh 90 lbs. to a bushel, up to as much as 140 lbs., but it is generally allowed that 112 lbs. is sufficient for a first-class cement, although from 100 to 105 may be considered as the average weight ordinarily sold by the best manufacturers; and if thoroughly pulverised, this is better for concrete building than one weighing 112 lbs. imperfectly pulverised. When cement is packed into sacks or casks at the works, it necessarily becomes compressed, and its particles also forced closer together in transit, and this should be borne in mind when a specification states that a certain proportion of cement and aggregate shall be used, and each to be measured at the time the materials are about to be mixed; but as cement is now usually sold by weight, with a guarantee given if required that it weighs a certain number of lbs. per bushel, it would perhaps be more

E

satisfactory to specify a certain amount by weight of
cement to a cubic yard of the aggregate, calculating
the requisite proportions from the makers' guarantee,
and which it would be easy to verify.

Portland cement possesses many advantages as a
matrix that other materials are deficient in; it does
not deteriorate by age if kept from wet or damp, it
does not expand or slake when mixed with water,
although allowing time to incorporate the ingredients
in sufficient quantity and deposit them in place
without setting, yet it hardens sufficiently fast to permit usual building operations in concrete to be carried
on with safety and more rapidity than could be
obtained with bricks and mortar. It resists damp
and withstands frost better than any other material;
it is hydraulic in character, setting in water equally
as well as in dry situations; sea-water can be employed for incorporating the ingredients without any
deleterious effects—in fact, it is stated by some that
salt-water tends to develop the hardening properties
of cement more than spring or rain water; and, as
far as experiments have yet been able to prove, it
crystallizes and increases in strength with age. Mr.
Grant's experiments have proved that Portland cement
two years after use is more than double the strength
it is when only a week has elapsed, as the following
table will show :—

The Results of 960 *experiments with Portland cement weighing*
    112 *lbs. to the imperial bushel gauged neat.*   *Years* 1862
    and 1863.   *John Grant, C.E.*

Age and time of immersion in water.	Cement neat, breaking strain.
One week	445 lbs.
One month	679 ,,
Three months	878 ,,
Six months	978 ,,
Nine months..........................	996 ,,
Twelve months.......................	1075 ,,

The cement tests were $1\frac{1}{2}$ inches square, and
subjected to tensile strain by means of an ingenious
instrument made specially for that purpose. This
and similar experiments with even more favourable
results, speak for themselves when Portland cement
concrete has to be compared with brick buildings for
strength, for it must be borne in mind that concrete
walls are, as a rule built very quickly, and are
subjected to extraordinary strains, from the fact that
often all joists or other timber ties are left out till the
walls have attained their full height, and that the
scaffolding on which the workmen are employed is
often suspended to the walls during their construction
and without any other form of support. But it may
be supposed that the strength of Portland cement
used neat, or without the admixture of any other
material, is a favourable way of testing its powers,
and which would not stand the same relative time tests
when mixed with an aggregate ; but the actual result
is otherwise, as the following experiment by Mr.
Grant will show. When mixed in the proportion of

1 of cement to 5 of sand, the breaking weights are at—

One month	21 lbs.
Three months	88½ ,,
Six months...........	95½ ,,
Twelve months	122 ,,

It may be assumed, therefore, that whereas neat cement at 12 months' old is about 150 per cent. stronger than at a month, cement concrete is nearly 600 per cent. stronger. Here, then, is a satisfactory proof of the vastly superior powers of Portland cement concrete to ordinary brickwork, or other forms of house-building, for if the walls can be erected (as they often are) in a month, without any timber ties and with the additional strain of scaffolding, workmen, and materials acting as a lever to force them apart, and yet be of sufficient strength, we have ample proof that although possessing at that time a margin over and above any future requirements, the walls will possess six times greater strength after a year has elapsed, and even then not have attained their ultimatum in that respect.

Another series of experiments made by Mr. Grant, as to the relative strengths of various descriptions of aggregates, and to test whether the greatest cohesive properties would be obtained by allowing concrete to harden in water, or in the atmosphere, showed that, as a rule, Portland cement concrete was strongest when air-dried, and that in point of strength the most suitable aggregates stood in order thus—

1. Portland stone.   3. Flints.   5. Granite.
2. Pottery.   4. Glass.   6. Ballast.
                7. Slag.

These were all necessarily crushed, ground, or broken to a suitable size. But this order sometimes varied, except in the case of Portland stone and pottery, which in all the experiments maintained their character. The actual force of compression, or weight, necessary to crush test blocks of concrete made from the materials named was as follows :—

Size of block, 6 in. by 6 in. by 6 in. Moulded Nov. 6, 1867; tested Nov. 6, 1868. Proportion of materials, 1 part of cement, by measure, to 8 parts of the aggregate, the latter being of a proper size, and the concrete blocks compressed.

	Kept in Air.	Kept in Water.
Portland Stone	33 Tons	29 Tons.
Pottery	22 ,,	23 ,,
Flints	17½ ,,	20 ,,
Glass	18 ,,	17½ ,,
Granite	19½ ,,	16 ,,
Ballast.........................	13¼ ,,	13½ ,,
Slag.........	19½ ,,	13½ ,,

The same proportion of materials, and the conditions in all respects similar to the foregoing experiment, but the concrete blocks not compressed, gave the following result :—

	Kept in Air.	Kept in Water.
Portland Stone	24½ Tons.	19½ Tons.
Pottery	18 ,,	18 ,,
Granite	14½ ,,	13½ ,,
Flints	14 ,,	12½ ,,
Glass	13½ ,,	11½ ,,
Ballast.........................	12½ ,,	11 ,,
Slag	14 ,,	9½ ,,

This experiment proved that where compression, or a gentle impingement can be practised, as in trenches for foundations, arches, floors, &c., it is an advantage; on the other hand, for ordinary wall building, no movable appliances can be made sufficiently strong to allow of this being done, without greater corresponding disadvantages; but where increased strength is absolutely necessary for some special purpose, as for piers, buttresses, &c., it can be obtained by increasing the proportion of cement: thus, 1 portion of cement and 6 of aggregate required the following crushing force, the remaining conditions being the same as in the two former experiments, the concrete not compressed :—

	Kept in Air.	Kept in Water.
Portland Stone	30 Tons.	23 Tons.
Pottery	24½ ,,	24 ,,
Granite	24½ ,,	15½ ,,
Slag..............................	20 ,,	19½ ,,
Ballast..........................	18¼ ,,	17 ,,
Glass	16½ ,,	17 ,,
Flints	15¼ ,,	15½ ,,

These results tend to show that the concretes which gain strength most in proportion to the cement employed, are composed of the following aggregates, as in following order :—

Granite,   Ballast,   Slag;

and that these, with an additional 25 per cent. of cement, gain 50 per cent. of strength.

The general result, as well be seen, of these experiments shows a necessity for the aggregate to possess a moderate amount of porosity ; the comparative weakness of ballast is probably due to the want of angularity, caused by attrition, and the smooth surface and non-absorbent properties of glass and flints would fully account for these two aggregates standing last in the order of strength ; but slag being moderately porous, of close texture, and capable of a large amount of compression in its natural state, should have stood higher than the actual result placed it.

For comparison with the strength of ordinary brickwork, test blocks were made with concrete made from cement and various aggregates, with blocks of brickwork cemented together, and with blocks of neat cement, and after being allowed to harden for a year, the result was as follows :—

	Dimensions.	Crushing Force.	Crushing Force per sq. inch.
1 Portland cement to 8 of Thames ballast	12 in. by 12 in. by 12 in.	Tons. $70\frac{1}{2}$	Lbs. 1,098
Bricks joined with half cement and half sand...... ..	$13\frac{1}{4}$ in. by 12 in. by 9 in.	$21\frac{1}{2}$	299
Neat Portland cement	12 in. by 12 in. by $12\frac{1}{2}$ in.	Not crushed with 100 tons.	

Scarcely any better proof positive could be desired

or obtained, of the vast superiority of concrete over brickwork in point of strength.

It has generally been assumed that sharp, coarse Thames, or other river sand, is the most fitting constituent for mixing with Portland cement for use as a mortar, or for stucco, or other purposes, where for obvious reasons neat cement cannot be employed ; but another experiment made by Mr. Grant proved that good pit sand was sometimes superior for the purpose (where strength was the consideration), for whereas test briquettes having a sectional area of $2\frac{1}{4}$ superficial inches, made of half cement and half Thames sand, broke, at the end of twelve months, with a tensile strain of 724 lbs., it required 815 lbs. to accomplish the same when half cement and half clean, sharp, pit sand was used ; and a still more interesting experiment made by Mr. Grant as to the fitness, strength, and peculiarities of various kinds of sands when used in conjunction with cement, the proportion being one of the latter to two of the former, and a tensile strain being employed, the samples having been immersed in water, gave a result as follows, per square inch :—*

* Grant—on the Strength of Cement.

TENSILE STRENGTH.		
At three weeks' immersion.	At three months' immersion.	
lbs.	lbs.	
269	355	Clay ballast, burnt and ground, of a pale brick-red colour, with a rough uneven grain, and containing a good deal of dust. Mortar very coherent, apt to shrink and crack; strength very uniform.
165	254	Portland stone dust, a mixture of roach and whitbed ; grains rough and irregular, quite clean. Mortar tolerably coherent; strength very uniform.
140	249	Sea sand, with roughish and uneven grain, chiefly silicious and quite clean. Mortar very short and coherent; strength of different samples somewhat variable.
108	248	A silicious pit sand, containing a number of minute shells and a small quantity of some orange colouring matter ; grains semi-transparent, brownish yellow, and of unequal size. Mortar rather short ; strength tolerably uniform.
60	193	Drifted sea-sand, with pure silicious, semi-transparent, and almost colourless grains, and quite clean. Mortar rather short; strength tolerably uniform.
94	175	A silicious pit sand, quite clean, with uniform, smooth, semi-transparent grains. Mortar somewhat short ; strength very uniform.
38	91	Smith's ashes, containing, however, a good deal of unburnt coal dust; grains rough and irregular. Mortar moderately coherent, set slowly, and apt to shrink and crack ; strength very uniform.

These figures might be somewhat anticipated from the result of the trial of aggregates suitable for concretes, when tested by compression,—crushed Portland stone, and brick, pottery, or clay ballast ground, taking the lead; and Portland stone dust, and ground brick ballast and pottery, standing first in order as the most suitable for making cement mortars, where strength is the object sought.

Another point may be noted as one of the results of the experiments already named, that where it may be essential to economize the consumption of cement, a choice of aggregates—if a choice exists—may allow a smaller amount of cement to be used, and yet secure a concrete of average strength; for, as already stated, 6 parts of ballast to 1 part of cement crushed at 18¼ tons; but 8 parts of pottery and 1 of cement required 18 tons to accomplish the same object. From these tables may be also learned the information, that the non-absorbent materials lose strength, when a less proportion of cement is used, in a much less degree than absorbent aggregates; thus at 10 to 1 not compressed (the blocks measuring 12 in. × 12 in. × 12 in.) granite made a concrete nearly equal to Portland stone, and flints equal to pottery, while glass was superior to slag. Although these experiments, as shown by the tables, vary a good deal in their results, yet they indicate with tolerable accuracy what are the best aggregates to use in conjunction with Portland cement; but concrete made in an ordinary manner, with ordinary workmen, probably would not withstand more than one-half the force applied by Mr. Grant, who selected cement of high quality, and, without doubt, procured aggregates free from any deleterious matter; yet even then the margin of strength over and above that which is usually necessary in ordinary buildings, is in excess of any reasonable requirements.

Portland cement should not be employed as a matrix for concrete for some time after being manufactured, although there is not so much danger to apprehend in this respect when employed as stucco for ordinary plastering, for concrete, as a rule, is not so compact and dense as cement mortar, and in ordinary buildings no large masses of concrete are necessary, and more freedom is consequently given the cement in case of any displacement or rupture of its particles from excess of lime in its manufacture, or from other causes. A simple test to enable anyone to judge whether Portland cement is fit for use or otherwise is to make up a small pat of neat cement mortar, and place the same in water; if after twenty-fours it is free from cracks, it may be safely employed; but if otherwise, it should not be used for several days, when another pat may be made in a similar way and again examined. The light cements are not so liable to rupture in setting as the heavy ones, which have a larger proportion of lime stone, or chalk, used in their manufacture, and the component parts of which are burnt at a very high temperature; but as cement, whether light or heavy, loses no portion of its strength by being kept a reasonable time prior to use, it is better under any circumstances, as before stated, not to use it too quickly after its receipt from the manufacturer.

It is a mistake often made, to suppose that when Portland cement, or cement concrete, has no perceptible sign of setting or hardening for some considerable time after mixing, that the former is of inferior quality, for as a rule it may be assumed to possess qualities of quite an opposite character. The

colour of Portland cement should be, when mixed with
water and allowed to set, a bluish grey; if it dries a
yellowish brown it is a sign of possessing too great a
proportion of clay, and is then probably quick setting
and of low specific gravity. After concrete walls have
been executed some weeks (should they not be coated
or covered with any material in the meantime), the
cement contained in the concrete, if of good quality,
changes in the process of crystallizing to a lighter
grey colour, and gives the walls an appearance
altogether different from those constructed with inferior cement, or unsuitable aggregate, or perhaps a
combination of both, as the concrete in this case
undergoes little or no change in colour, but retains a
dull sombre hue. If a small quantity of cement be
placed in a vessel with clean water, and after well
stirring allowed to settle for a few minutes, and the
liquid portion drained off, sometimes a quantity
(more or less) of coarse, dark, insoluble grains remain
at the bottom—these are either portions of the cement
not ground sufficiently fine in their passage between
the mill stones, or, slag dust, or similar matters of
adulteration added to increase the weight, and thereby
either reduce the selling price of the cement, or
increase the manufacturer's profits. In the former
case, the particles of cement have little or no cohesive
properties, but are inert in their action to a great
extent. In the latter instance, although productive
of no actual injury to the work, they reduce the
strength of the cement in a similar manner as if sand
were added, and for which the buyer necessarily pays
the value of cement; ground hydraulic lime is sometimes added for the purpose of adulteration, but its

low specific gravity prevents its being used to any great extent. These practices, however, are not usual with any of the well-known manufacturers, whose interest is too great in producing an article which should stand fair ordinary tests, and the demand for which commands a ready and continuous sale at a remunerative price. But it is still desirable on works of any pretensions, or where cement is in common

Fig. 1.

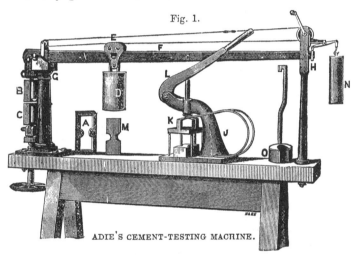

ADIE'S CEMENT-TESTING MACHINE.

use, to employ some mechanical means of testing its properties, and for this purpose cement-testing machines have been invented; the two most generally used are known as "Adie's" and "Michéle's." Adie's (Fig. 1) was selected, and in fact invented, almost expressly to meet the requirements of the Metropolitan Board of Works for the experiments made by Mr. Grant, and is a modification of the lever balance. The test cement is mixed with as little water as possible, consistent with perfect homogeneity, and is then deposited in the mould A *, which has previously had a plate of iron M placed therein for the cement to rest

Fig. 2.

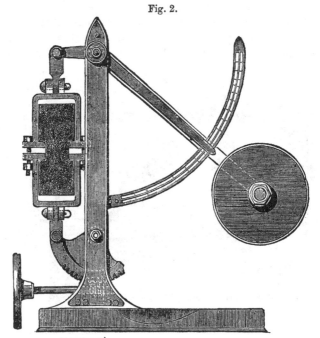

MICHELE'S CEMENT-TESTING MACHINE.

upon, the mould is filled with cement, and the latter scraped off flush, or fair with it ; when sufficient hardness has been obtained, the mould is placed under the cage K, and by depressing the lever L the brick is

* Mr. Adie is now making the mould A of an improved form (the result of many careful experiments by Mr. Grant). which enables the cage and press L K J to be dispensed with.

forced gently from the mould and then placed in water, or in the air, according to the nature of the experiment, and until such time as is decided upon to test its strength, when it is placed between the clip B and C; the wheel I, which is attached to a screw in connection with the clips, is then turned round till the latter clasp the brick with sufficient force to raise the end of the beam at H up to the pulley, the main weight E being kept near the pillar G. The main weight E is then by means of the lever cord attached to it wound gently along the beam until fracture occurs. A series of figures on the beam indicate the amount in pounds at which the breaking strain takes place. The cost of Adie's machine varies with the nature of the material it is made of, and the number of moulds, &c., supplied with it. The machine sold at £20 is sufficient for all ordinary purposes, being capable of testing blocks up to 1,100 lbs. tensile strain. Mr. Adie has also a still smaller machine, the cost of which is £16, for testing samples of 1-inch sectional area, and combining the power of an ordinary steel-yard. Michéle's machine (Fig. 2) is constructed on quite a different plan, the principle, however, being the same—of employing a weighted lever capable of exercising a very gentle increasing strain. The test cement is made and moulded in a similar way to Adie's; the block to be tested is placed in the clips, the handle is then turned, which raises the weighted lever by exerting a pull on its short end through the medium of the cement block or brick. When the leverage is so increased as to exert a force too great for the cement to sustain, it breaks and the lever falls, leaving the index pointer at the spot to which it has been raised.   The arc along which the pointer moves is graduated to show the number of pounds of tensile strain applied. A simple arrangement when the cement block breaks prevents the lever falling more than half an inch. Michéle's machine costs from £20 to £30, according to the amount of strain it is required to be capable of exercising. Both the cement testers described, as will be seen, are efficient and simple in action, and one or other should be in the hands of all users of cement and intending builders in concrete.

CHAPTER IV.

PORTLAND CEMENT is used as a matrix for probably four-fifths of all the concrete buildings executed ; but ordinary ground lime, with the addition of a small proportion of sulphate of lime, has been patented and somewhat extensively used as a matrix under the name of " selenitic cement," or as it was first called, " selenitic lime."

Some years back, General Scott patented an article designated " Scott's Cement," and which consisted of ordinary lime made in the usual manner, but exposed to the action of sulphur fumes, the sulphur being burnt in vessels placed beneath the lime-stone during calcination ; the lime was afterwards ground fine, and used chiefly for plastering purposes, as it then possessed the important property of not slaking, or swelling, when water was added, but could be used in the same manner as a cement, and applied without any delay. The disadvantage it possessed was through the great difficulty experienced in submitting the lime contained in the kiln in an equal degree to the sulphur fumes, whereby some portions of the lime became overcharged with the same, while others remained

F

untouched. General Scott, in a lecture to the Architectural Association in 1871, states how the idea
occurred to him of adding sulphate of lime to ordinary lime to prevent it from slaking when being
converted into mortar, thus: "While making some
experiments with the Plymouth limes, in the course of
which I found an excellent hydraulic lime among some
beds generally rejected for burning, I discovered
certain curious effects produced by burning the stone
in a dull fire,—I found, in fact, that the lime burned
in this way, in lieu of slaking and heating, as I should
have anticipated, when reduced mechanically to a fine
powder, and treated with water, set into a solid mass.
I consulted Professor Faraday respecting the phe-
nomenon, and he eventually came to the conclusion
that this change in its behaviour was due to the
formation of some portion of "subcarbonate of lime."
After many experiments, I ascertained that this action
was really due to the presence of a small quantity of
sulphate of lime, resulting from the oxidation of sul-
phurous acid, arising from the fuel, which had been
mixed in along with the lime in burning."

In 1870, selenitic cement, which consists, as in
Scott's cement, of the addition of a small quantity of
sulphate of lime to ordinary lime, was patented, and
instead of burning sulphur to effect the desired purpose, sulphate of lime in the form of green vitriol,
gypsum, or plaster-of-Paris, is added, either at the
time of grinding, or when required for use.

Why it is called cement when it consists almost
wholly of ordinary lime is, according to General Scott,
because the former name belongs to any cementitious
material which unites with water and passes into the

hydrate in one operation, whereas a lime slakes first
and combines with silitic acid subsequently. The
term "selenitic" is derived from the word "selenite,"
the chemical name for gypsum—the stone from which
plaster-of-Paris is made by burning, and which latter
is added to the lime while being ground in the mill,
or mixed with water at the time of use. For concrete purposes it is better to have the selenitic cement
ready prepared, and so to avoid possibility of any
portion being used as a matrix without the necessary
sulphate, which would probably slake and cause
disintegration of the concrete.

The Selenitic Company's instructions for making
concrete are these: "Throw into the pan of the
edge runner two or three parts of water (according to
the moisture in the sand); to the first of which 2 lbs.
of plaster is to be added, and gradually introduce a
bushel of prepared lime; continue grinding until the
whole is reduced to a creamy paste, and then put in
one bushel of sand. The mortar thus prepared to be
turned on a board with from five to six parts of
ballast." The addition of the plaster is assuming that
it has not been already mixed with the lime in the
process of manufacture, but even then precaution
should be taken to ascertain if sufficient has been
added, which can be readily done by putting a small
quantity in water; if the latter becomes warm, some
plaster should be mixed with the water used for
forming the concrete. Another method of determining
its suitability is to make a small pat of neat lime
mortar; if the latter cracks in drying, that also is
a proof of more plaster being needed; but this addition of plaster should not, however, be overdone,

for no good would result therefrom—possibly the reverse.

It will be seen that selenitic mortar is in reality composed of the same materials as the ordinary gauged mortar used by plasterers; but the latter allow the lime to slake and cool first, adding the plaster after it is made into a mortar, to create quick setting properties, and not to increase the strength of adhesion; but by the selenitic method the loss of power sustained by the usual way of mortar-making is avoided, and all the valuable properties of the lime economized.

But although any lime can be made into a selenitic cement by the patented process, it does not follow that they are all of equal strength and quality, or equally adapted for concrete building; the selenitic cement made from lime of only moderate hydraulic properties is scarcely suitable for ordinary concrete walling, and certainly not for buildings in damp situations, for foundations, or any purpose whatever, where it cannot be protected by a coat of cement, or otherwise shielded from moisture. For concrete walling, selenitic cement should be made from one or other of the well-known hydraulic limes, as the Rugby, Barrow, or Aberthaw, and concrete with selenitic cement as a matrix made from either of these or other well-known powerful limes, will make substantial walls, especially if a small proportion of Portland cement be added. The following table of the relative strengths of various limes and cements is from experiments made by Mr. Kirkaldy, with the view of testing the capabilities of selenitic cement for the Selenitic Cement Company.*

* 1, Great College Street, Westminster.

But it must be noted that it does not give the normal strength of Portland and Roman cements, for comparison with limes in such way that a correct judgment may be arrived at, as no two test blocks were of the same age, and the figures altogether seem a puzzle, for Dolgoch selenitic lime appears to be of about equal strength whether mixed with three, four, or five times its volume of sand, but when increased to six times, over 33 per cent. additional strength is at once gained.

Table showing the relative breaking weights in pounds of briquettes, having a sectional area at the neck of two and a quarter superficial inches.

Nature of lime or cement.	Age in days when fractured.	Composition of Mortar.				Pounds per superficial inch, with 6 parts of sand.	Remarks.
		3 of sand to 1 of cement or lime.	4 of sand to 1 of cement or lime.	5 of sand to 1 of cement or lime.	6 of sand to 1 of cement or lime.		
Roman cement	132	232	250	124	89	40	
Portland cement	167	,,	206	149	113½	50	
White chalk lime	164	67½	,,	,,	,,	,,	
,, selenitic .	161	63	58	78	72¼	32	
Burham lime, selenitic	165	,,	,,	170	210	93 ⎫	Good Medway
,, ,, ,,	234	,,	340	,,	,,	,, ⎬	grey lime sold by
,, ,, ,,	161	255	,,	,,	250	111 ⎭	Messrs. Lee.
Halkin lime, selenitic .	76	128½	197	99	111	49	Good hydraulic
Dolgoch lime, selenitic	62	155	156½	157	206½	92	Very hydraulic

In this table the Burham lime appears to stand first

in quality, and the statement of the extraordinary high strain it withstood would, except on so good an authority as Mr. Kirkaldy, have been open to doubt; and if selenitic lime concrete would show the same comparative result when tested with Portland cement concrete, the latter material must then certainly make way in favour of the former for building purposes. On the other hand, Mr. Grant stated at a meeting of the Institute of Civil Engineers, that he made a number of briquettes, or moulds, of selenitic cement, but they were so weak as not to bear winding up in the machine, and that in an arch of one of the largest sewers, he found it inferior in quality to ordinary blue lias lime.

But the strength of cohesion in lime is not in proportion to its strength of adhesion, whereas with Portland cement the case is quite different; for it is, as has been shown, strongest when used or mixed neat, that is, without the admixture of sand or any other solid matter; but selenitic lime is stronger when six parts by measure of sand are incorporated with it than when three only are employed, and the latter would probably be treble the strength of neat lime mortar.

The Selenitic Company state that a given quantity by measure of their cement when mixed with from six to seven times the same quantity of Thames ballast or burnt clay, will be almost equal in strength to the same quantity by measure of Portland cement when mixed in like proportions; this, however, is not borne out in practice, and there is no doubt that selenitic cement made from the best hydraulic lime forms a good concrete for building purposes at a somewhat less cost, but of inferior strength to Portland cement. The Selenitic Company also say, that by the addition of one-fourth part of the best Portland cement with the selenitic cement, the results are improved for concrete walling. This would appear to be an anomaly, because if the former is of equal strength to the latter when mixed with six or seven times its bulk of an aggregate, how can it be improved by any substance no better than itself?

The cost of selenitic cement varies with the quality of the lime it is made from, but averages probably about 30s. per ton at the works, and measures about thirty bushels, as against Portland cement at 40s. per ton, of twenty to twenty-two bushels; but this comparison of cost must be regulated by the amount of carriage in either case, and also, if any Portland cement is added to the lime; but with the present moderate price of Portland, it is doubtful whether selenitic cement will come into very general use as a matrix in concrete construction.

Great care is exercised in preparing it, so that no coarsely-ground, dangerous particles are allowed to pass through the mill stones, and it can therefore be safely employed without fear of causing disintegration; but it will not retain its strength when exposed to the air to an almost unlimited time, as is the case with Portland cement; in other respects, it has no great disadvantages compared with the latter material, for it makes walls impervious to moisture, and sets sufficiently fast for ordinary buildings (although not so rapid as cement), and can be treated in all respects similar. As plaster of Paris is soluble in water, it has been supposed it would, when mixed with ordinary hydraulic lime, tend to diminish its hydraulicity; but it does not appear to have this effect, or, at any rate, but very little—the quantity employed, 5 per cent., not being sufficient probably to alter the character of the powerful lias limes.

Good concrete may be made from any of the ordinary hydraulic limes, but great care is necessary in their use, and it is indispensable that the lime should be ground fine. When Portland cement—about two years since—through the excessive cost of fuel, attained a very high price, many substitutes were experimented upon, and some excellent results obtained from the well-known blue lias limes. The great difficulty experienced is to prevent the expansion consequent on slaking, and as hydraulic lime is much slower in its action in this respect than the rich, chalk limes, the difficulty is thereby increased.

The general method recommended is to cover as much lime made into a conical heap as is required for a day's consumption, with damp sand; the moisture from the latter is converted into steam by the heat given out from the lime in slaking, and the whole of the latter becomes an hydrate; from twelve to twenty-four hours is necessary for the lime to be covered with the sand, and it should then be used as quickly as possible, or it will deteriorate in strength. Another method is to add a small proportion of moist sand to the lime, well incorporate the two, and occasionally stir, or move, till the lime has abstracted the moisture, as in the former case; the objection to this plan is, that if the particles of sand are not equally diffused, a large amount of moisture may be concentrated at certain places, and the formation of knobs, or balls of mortar, will be the result. But if the lime is emptied from the sacks in which it is sent from the grinding mill on to a wood floor, and occasionally turned with a shovel, or even with a rake, so that air is admitted and diffused through it, and allowed to remain thus for a week, or longer, before employed as a matrix, no other treatment is necessary, the dampness in the atmosphere having done all that is required ; this is called "air slaking," but although air should be transmitted through the lime as much as possible, a dry floor is very necessary to deposit it on, and, of course, be thoroughly protected from rain or moisture.

If sieved through a fine sieve, in most hydraulic ground limes an amount, more or less, of coarse particles will be found ; and if these prevail to too great an extent, they are evidence of the lime itself not being thoroughly ground, but are not all necessarily unground portions of lime, as the limestone itself often contains granular, sandy substances, which do not calcine, and a portion of the silica which forms so large a component part of the limestone is granular ; if these coarse grains be mixed with water, the result, whether lime or otherwise, is quickly determined. Lias lime concrete is improved in strength by the addition of Portland cement—one part of cement, two of lime, and twenty-one of aggregate, by measure, make a first-class concrete. The Bridgewater blue lias lime will compare with almost if not all that is produced for cheapness and strength, as it is sold delivered into railway trucks at the works in bags, at from 12s. to 13s. 6d. per ton of thirty bushels, and if employed in the proportions named, and treated in the way described (with the addition of the cement),

is but little inferior to Portland cement for concrete.
Large blocks of buildings have been erected with
concrete made from this formula, the result being
everything that could be desired. In using blue lias
lime, it should not be treated in the manner described
for selenitic cement, but in every way similar to Portland cement, and which will be more fully described
hereafter.

CHAPTER V.

THE most important part of the plant required for monolithic concrete building is necessarily the mould, or frame, into which the concrete is deposited in a semi-liquid state—and assuming the aggregate is obtainable of the necessary size without the intervention of a stone-crusher, but few other appliances in an ordinary way are wanted. But, although of a simple character, these are essential, and should always be employed, for especially with concrete construction does the proverb figuratively apply, "The strength of a chain depends upon its weakest link."

The first point to be observed is, that the aggregate should be deposited in a clean place—on old planks, scaffold boards, or anything whatever that will prevent dirt becoming intermixed with it; and, obviously, as near as circumstances will permit, to where it will be required for use. The concrete itself should be mixed on boards, which can either be laid down singly, or, better still, ledged together on the under side. In the latter case, the ledged boards should be in two portions, so that when placed together they measure not less than nine, or more than twelve feet square;

if less than nine feet, there is not space to mix a sufficient quantity of concrete thereon at one time, and if more than twelve feet, the boards are cumbersome and unwieldy, besides being of useless dimensions ; the medium is a suitable size, and the motive for making them in two portions is for convenience of removal, and to enable them to be passed through any ordinary doorway. Each of the "ledges" to which the boards are nailed should project six inches on one side, so that when the two halves of the mixing-boards are laid together for use, the projecting portions of ledges of one half form a bearing for the corresponding half, and make the whole substantial and firm. (Fig. 3.)

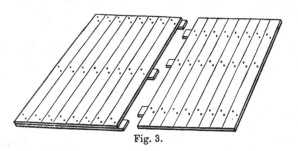

Fig. 3.

The boards need not be more than an inch thick, and the "ledges" should measure about 5 in. × 1 in. ; the nails should be driven through the boards, be long enough to pass through ledges as well, and to turn down, or "clinch," otherwise the effects of the wet concrete and the rays of the sun combined may cause the boards to warp and form an uneven surface; the nails should also be made to enter, or "punched"

half-way through the boards, or, as the latter wear
thinner from the effects of the workmen's shovels, the
heads of the nails will remain above the wood, and
impede the progress of the work. The mixing-boards
should be laid level, or nearly so, except the aggregate
requires washing, when they may be slightly inclined,
as previously stated, to allow the water to pass away.
As many mixing-boards may be employed as the
amount of work to be executed may justify, but it is
unprofitable to have too few, as the labour of carrying
concrete in buckets to any distance is considerable.
A two-pronged hook or rake should also be provided,
to assist in incorporating the ingredients ; this must
be similar to a plasterer's hair-hook, except somewhat
stronger, and have a handle seven to eight feet in
length. An ordinary watering-pot, to hold a large-
sized pail of water, is necessary, and having a fine
" rose " attached thereto, so that the water may be
distributed gently, because, if thrown, or poured on
with force, the finely-ground lime, or cement—as the
case may be—instead of being merely moistened, will
be washed away from portions of the aggregate ; the
rose should be soldered or otherwise permanently
fixed to the nozzle of the watering-pot, for no matter
how careful a supervision is exercised, the rose will be
discarded whenever an opportunity offers, or the
holes in same will be increased in size, to enable
the water to run quickly out. The better plan is
to have a purpose-made rose of strong sheet-iron ;
and another advantage in having one to distribute a
fine spray of water is, that, if the latter be dirty, it
rapidly closes the perforations, and to use clean
water becomes therefore unavoidable. Dirty water

for making concrete, it must be obvious to everyone, is as fatal for sound work as a dirty aggregate or an inferior matrix.

The measuring-boxes are important items, for on the regular and proper proportion of materials much of the character of concrete depends. The most common forms in use, and the simplest, are as shown by Figs. 4 and 5.

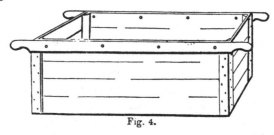

Fig. 4.

Fig. 4 is a rough wooden box without a bottom, having strengthening pieces of wood, forming also ledges, fixed at each external angle ; and angle pieces of wood, fixed at each inside corner, also serve to strengthen it. Pieces of wood bolted, or otherwise fastened to each side at top, serve as handles. This box is for measuring the aggregate, and the smaller one (Fig. 5), which is of the same length and width as

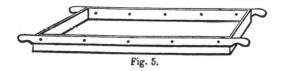

Fig. 5.

the larger, is for measuring the matrix. The relative dimensions, of course, depend upon the proportions to

be used of each material, and the quantity proposed
to be mixed at one time; but a convenient size for
many purposes is for the larger measure to hold fourteen bushels, which is exactly two-thirds of a cubic
yard. The dimensions in this case would be four feet
long, three feet wide, and one and a half feet deep
(internal measurements), equal to eighteen cube feet;
and if the proportion of matrix is to be as 1 to 7,
which would be equal to just two bushels, the smaller
measure must be two and five-eighths inches and a
sixteenth of an inch over in depth. The method of
measuring is properly done thus: The larger measure
is placed on the mixing-board at about two feet from
one of its sides (but in the opposite direction must be
in the centre), the aggregate is placed inside,
and with a wooden straight-edge levelled fairly with
the four top edges of the box; the smaller measure is
then placed on the larger, and the cement, if cement
be used, put in and struck off in a similar way to the
aggregate. The disadvantage of this plan to the contractor—if it be contract work—is that a certain quantity of cement over and above the specified proportion
fills the interstices of the aggregate, and if the latter
be somewhat coarse the amount of cement which is
thus added to that stipulated is considerable. Sometimes cement is delivered in two bushel sacks, and is
emptied on the aggregate without the intervention of
a proper measure; if it is certain the bags contain
two bushels, and are each and all alike, this is as good
a plan as any, but this cannot always be relied upon,
and a better and fairer way than either of the above
methods described would be to fill and weigh each
sack of cement required for the day's work, the weight

being regulated according to the specification or the formula agreed upon.

Another principle of measuring the materials is by making the box the proper dimensions to hold both aggregate and matrix; then, by using a wood straight-edge notched at each end, the aggregate can be levelled at any required distance below the top, as section (Fig. 6), the space between which and the

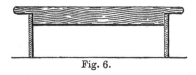

Fig. 6.

aggregate is that necessary for the matrix, and which is placed therein and struck off with the same straight-edge reversed, as section (Fig. 7). The advantages of

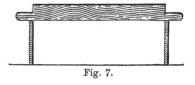

Fig. 7.

this method are that only one measuring box is wanted, and the proportions can be varied if required by merely regulating the depth of the notches in the straight-edge. A third way is sometimes adopted of having the box the necessary size to hold the matrix only, and which is filled with the aggregate as many times as the proportion of the same is to be to the one measure of cement. In this case the measuring-box must obviously have a bottom; the plan has some advantages, for the proportions can easily be varied, and no extra cement is gone amongst the aggregate, as would be by the two first methods described. When more than one matrix is used, as for instance lime and cement, it is usual to place the smallest in proportion on the top of the larger, as they are more likely to be thoroughly intermixed by so doing.

But any of the methods named may lead to some dissatisfaction in contract work on the part of either contractor or employer, or both, and doubts as to the specified quantity of the matrix being used may arise. The most satisfactory way, therefore, would be to employ a bottomless measure, as Fig. 4, for the aggregate, and for the matrix a cylindrical measure, the same as used for corn, but adapted for the exact amount of cement or lime, or one for each if they are both employed, would be found most convenient.

The aggregate having been placed in the larger measure, the cement by one or other of the ways described deposited on the aggregate, the measure or measures are removed by a workman at each corner lifting them perpendicularly by the handles, leaving the materials on the boards in the form roughly of a cone or pyramid. Two men, one on each side of the heap, then begin to throw the materials to the opposite side of the mixing board; another, standing by the heap now forming, further incorporates the ingredients by raking them backwards and forwards with the hook until the heap has been re-formed. The same process is then repeated, and by which the materials are replaced where at first deposited. Having now been turned over and raked "twice dry," the operation is again gone through in the same way, but with the help of a fourth man, who stands behind the heap and adds the water from the watering-pot to such portion as the two men who are shovelling are immediately about to remove. This makes three times turned and raked. Once more repeated finishes the process, and the concrete is then ready to be deposited where required, after having been turned over and raked twice dry, once during the process of watering, and once after.

It is essential that certain precautions be observed in mixing the concrete:—

1. The water should be added to that portion of the materials the two "shovellers" are working upon— not to the mass indiscriminately, as so doing would cause the cement to have time to sink through the interstices of the aggregate previous to attaining partial solidity.

2. Water should be added—as much as needed— during the third turning, not afterwards.

3. The amount of water applied must be regulated according to the purpose for which the concrete is intended. For foundations, arches, &c., where impingement can be practised, only as much as to cause slight cohesion between the materials is necessary; but for walls between frames, and similar objects, it must be in a kind of semi-liquid condition.

4. The "shovellers" must turn the concrete completely over when in the act of casting it from one heap to another—not take it up in the shovels—and deposit without changing the position of the ingredients.

5. If the aggregate is porous, say Bathstone chippings or crushed bricks, it is better, especially in hot weather, to well water it, and allow time for

absorption previous to use, or the aggregate will have taken away a portion of the water required for hydration. (If a dry brick be washed with liquid cement and exposed to the sun's rays, the cement, when dry, will have become an impalpable powder, having no adhesive properties.)

6. Not less than four men should be employed for mixing, nor will the variable nature of the aggregate employed, or other circumstances, justify any departure from the method of making concrete as described.

CHAPTER VI.

FRAMES AND APPLIANCES FOR DEFINING CONCRETE
WALLS.

THE simplest way of forming a mould or cavity, in which the concrete can be deposited in a soft state, is obviously by means of posts fixed in the ground, and on both sides of the intended walls, and nailing thereto rough boards or slabs. The posts must be connected together by bolts, ties, or some similar means, and the boards cannot be released until the whole of the temporary appliances are removed. This primitive form of defining the walls is still practised for constructing the clay, mud, or cob buildings in some parts of England.

Mr. Scott Burn, in a paper read to the Bath and West of England Society in 1869, suggests a somewhat similar plan, but proposes to make the boards into movable ledged panels, and securing them to the posts by angle irons, or plates and bolts ; at all corners or angles he would erect posts seven inches square, and the full height of the intended building, and—although not described it is assumed—intermediate posts of smaller dimensions, but of the same length or height. It must be quite evident to anyone who would calculate the cost of these kind of appliances that the first outlay would equal that of the most costly portable apparatus, and the application and means of adaptation for the different varieties of buildings they might eventually be needed for, must be equally as clear to anyone who has practised concrete building, would be anything but satisfactory or economical. For a straight low wall of moderate length, where no precision was necessary, this way of concrete building would answer fairly well, but with erections of various sizes and intricate plans, or with other obstacles to deal with, the cost would be as much as the walls themselves.

But the invention of movable appliances caused, to all intents and purposes, this method of forming a mould or matrix to be abandoned, and although there have been a considerable number of patents obtained for concrete wall building, including those by J. Tall, J. B. Tall, Drake, Osborn, Lish, Lythgoe and Thornton, Corpe, Potter, Macleod, Broughton, Brannon, and some others,—as is usual with most patented inventions, some have been found good in theory and wanting in practice, while others, through some unforeseen defect, have gone out of use, and the patent therefore abandoned.

Mr. Joseph Tall's, as before stated (Fig. 8), was the first movable apparatus introduced to public notice, and which gained considerable reputation from various causes, not the least being its adoption by the then Emperor of the French for the erection of a large number of workmen's dwellings in Paris, and the award of a gold medal to the Emperor at the Paris Exhibition of 1867 for the houses built by him.

The appliances consist of wood standards six to ten feet high, employed in pairs, one on either side of the

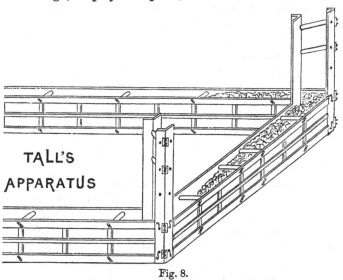

TALL'S
APPARATUS

Fig. 8.

walls to be built, and certain distances apart ; these are connected together with movable bolts and thumb-screws, the thickness of the walls being accurately determined by cores, or taper pieces of wood of the same length that the walls are required to be in thick-ness ; the bolts pass through these cores, and the latter are thereby prevented from being disturbed, and the bolts also from becoming fixed in the concrete ; when necessary to remove the standards, the thumbscrews are taken off, the bolts withdrawn, and the cores, being taper, are easily driven out. The movable panels or frames are 18 inches deep and various lengths, as may be required, and which are also secured with

thumbscrews and bolts to the standards. Holes in
the latter equally distant apart admit of the panels
being raised and fastened thereto as the work proceeds,
till the top of the standards is reached; these are then
removed and refixed by bolts passing through their
two bottom holes and through the top holes left in the
walling already executed. As this process can be
repeated *ad infinitum*, any height of walling can be
reached with the same amount of apparatus. For the
corners of walls angle posts are used, and the end
inside panels are necessarily the thickness of the wall
less in length than the outer ones. Brackets secured
to the walls with the same bolts used for the apparatus, serve as scaffold-bearers; and flue cores or
cylinders built in the chimneys, and withdrawn and
re-arranged as the work proceeds, form ventilating or
smoke flues. Various contrivances for constructing
chimney breasts and other projections also form a
portion of this patent.

Osborne's apparatus (Fig. 9) is similar in many
respects to Tall's, but the standards and movable
panels are fastened in their proper position by bolts
and rotary studs instead of thumbscrews, and it differs
also in some other minor points.

The patentee claims superiority for his apparatus
for a variety of reasons, but principally that the weight
or strain of the scaffold and the load it has to carry
may, by adjustable brackets, be distributed over
the greater portion of the entire area of the walls
already built, instead of the top portion only and
that projections may be formed for cornices, string
courses, &c., provision being made for leaving any
size aperture between the movable panels; but of

necessity the standards cannot be removed, and this
assumed advantage is therefore almost valueless.

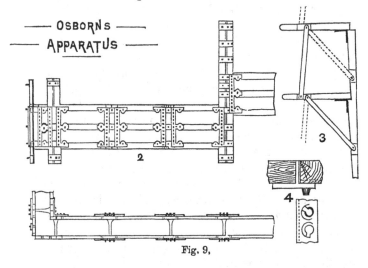

Fig. 9,

Fig. 1 is a plan of this apparatus ; Fig. 2, an elevation showing panels fixed at different heights ; Fig. 3,
brackets to be fixed to walls to carry scaffold for workmen ; and Fig. 4, larger scale plan and elevation of
the rotary stud fastenings.

Drake's apparatus (Fig. 10) differs materially from
the two described, the standards, instead of being of
wood, are channel iron, the flanges of which when fixed
being turned towards the walls about to be built ;
mortices through these flanges every two feet apart
admit of bars of iron about 1 in. wide and $\frac{1}{4}$ in. thick
passing through them ; these bars, which are called
"wall gauges," are pierced with holes $1\frac{1}{2}$ in. from
centre to centre, to admit iron pins, which pass at

the same time through small webs of iron riveted to
the standards, and regulate the thickness of the walls

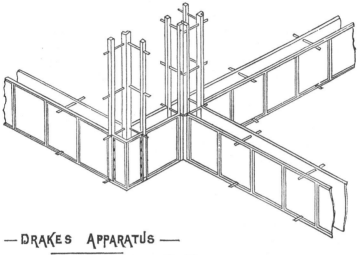

—DRAKES APPARATUS —

Fig. 10.

to be built, the length of the gauges and the extreme
distance of the pin-holes in them forming the limit
in this respect ; but, as a rule, they are supplied to
construct walls four and fourteen inches in thickness,
and including every intermediate 1½ inches. The
panels of sheet-iron, riveted on a framework of angle
and T-iron, are attached by malleable iron clips, and
pins or bolts passing through the latter and also
through the side flanges of the standards ; these
panels are made two feet in width, and consequently
admit of two feet in height of walling being constructed at one time. At intermediate distances between the standards, the same kind of wall gauges

are used, passing through the top and bottom angle
irons of the panels. For corners of walls angle-
plates are employed, which are secured to the end
standards, but as they would obviously only serve for
walls of one certain thickness, there are "adding"
pieces belonging to them, so that when the thickness
is reduced, the adding piece is fastened to the inside,
and when increased, to the outside angle-plate. Projecting plates for constructing chimney breasts and
other irregularities, telescopic plates for attaching to
ordinary panels to make up the exact length of walling required, and collapsing flue-cores which admit of
being withdrawn from the concrete when the latter
has set, form other portions of the apparatus. The
scaffold brackets are made of angle-iron, and temporarily fastened to the uprights by bolts and nuts.

Potter's appliances (Fig. 11) consist of T-iron standards, provided with a foot or shoe at bottom, by which
the standards are fixed to stumps driven in the ground,
or to cill-pieces, floor-joists, &c. The standards are
placed in pairs, one on either side of the intended
walls and connected together at suitable intervals in
their length by flat iron wall gauges, having a
shoulder or lug at one end, and a series of holes an
inch apart.

These wall gauges are passed through corresponding mortices made in the flange and web of the
standards, and are secured by pins or bolts passed
also through holes in the web of the standards and
through one or other of the holes in the tie bars,
according to the desired thickness of the wall, which
can be thus regulated as required.

a Flat iron bars are riveted to the T-iron uprights,

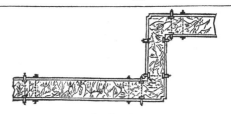

PLAN

*Slating Batten as a
Temporary Tie.*

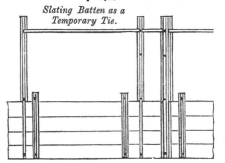

ELEVATION.

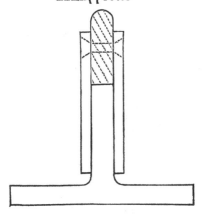

FULL SIZE OF STANDARD
Fig. 11.

one on either side of the web, the said bars being of greater width than the web, so as to strengthen the uprights and leave a groove in the edge for inserting plate-irons for projecting panels and other similar purposes. Common rough boards are placed against the flanges of the T uprights to make the mould, into which the concrete is filled to form the wall, and are retained in position by wall-ties, consisting of small iron cramps placed against the boards and connected by tie-bars, made adjustable by means cf pins or bolts, as before described.

Appliances for making circular or egg-shaped underground tanks are also included in this patent, and are described therein as follows : " In construct- " ing circular tanks I use L- and T-iron ribs, bolted " together in sections, which then form the inner line " of the tank or well, and the concrete is kept in place " by rough boards placed in the recesses of the T- and " L-iron ribs. These ribs can be placed any convenient " distance apart, and when the bolts connecting them " are withdrawn the framework is liberated. Oval or " egg-shaped tanks, whose lesser diameter is the same " as the diameter of the circular tank, but whose " greater diameter may exceed the diameter of the cir- " cular tank to any required extent, can be constructed " by using extra straight lengths of framework-iron " connected to the semicircular ends of bolts and nuts."

Lish's appliances are made of iron, and differ very considerably from any other, but the complicated system adopted would prevent its use in any building where cost was a consideration, nor is it possible to give any brief description of its construction, the Patent Office specification alone being accompanied with ninety-seven separate illustrations and sections necessary to explain the principle.

As most other appliances are in general character a good deal similar to those illustrated, it is unnecessary to describe them, but they are nearly all composed of two principal portions, viz. the standards, which are any required height, and are rigidly fixed, and on which the accuracy of the walls largely depends, and the corresponding panels, or frames, which are attached thereto, and serve to confine the semi-liquid concrete until it has become sufficiently consolidated to allow their removal; and the principle of construction is nearly the same in all, viz. building the walls the full height of the standards, then removing the latter and refixing them at the top of the walling already executed, and so on till the full height of the building is attained.

The disadvantage that nearly all concrete appliances possess, and for which there seems no simple remedy, is that owing to the standards being fixed in position close to the faces or sides of walls, considerable difficulty ensues if projecting cornices of stone or other materials require to be built in as the work proceeds; for although there would be none, provided the standards could be so adjusted as to move forward and backward from the face of the walls, the want has not yet been surmounted in a practicable way. Lish's, and one or two others, have been designed to meet this difficulty, but have not achieved success. But the panels or frames are more easily dealt with in this respect, and therefore cornices intended to be cemented and requiring only a core formed in the wall itself, cause little or no trouble, as the break or void in every

case where a standard is fixed, can easily be made
good after it is removed, but with a stone cornice this
could not be permitted, and special means have to be
adopted, therefore, to meet the difficulty.

Even the insertion of brick bond and terra-cotta
ornaments in the walls as they are built, cause a good
deal of extra labour, and this accounts in a great
measure for the discrepancies in estimates and
tenders for concrete buildings, as the more irregular
the plan and the greater the amount of projections, or
brick, stone, or terra cotta insertions in the form of
string courses, bands, corbels, mouldings, &c. the
larger in proportion will be the cost of the concrete
portion of walling, in many cases exceeding that of
ordinary brickwork.

Although made panels or frames to attach to the
standards for enclosing the concrete—as in Tall's,
Drake's, and Osborne's appliances—would be, no
doubt, better than any others if they were required
for buildings all of one shape, size, and internal
wall arrangement, yet, where no two may be alike,
considerable difficulty arises in adapting them to walls
of various lengths and thicknesses. To remedy this,
short panels or frames from three inches to two feet
in length, as described previously, and called " adding
pieces," " lengthening panels," " telescopic plates,"
and other names, are bolted to the nearest stock
length of panel at hand, to make as near as can be
obtained the length required, but at best they are
clumsy contrivances that unsatisfactorily answer the
purpose.

It is evident that the weight of sufficient standards,
panels, adding pieces, &c. &c., for an ordinary house,

is very considerable, and to remedy this and avoid the difficulty experienced in building walls of various sizes, ordinary rough boards of any thickness are employed (but inch preferred), and which can be cut to the lengths the buildings necessitate, and afterwards used as common flooring or roof-boarding, or even for fencing materials, assuming they are not again required for concrete construction. The doubt whether the boards would not become crooked from the effects of the wet concrete on one side, and the rays of the sun on the other side, has been solved by reversing the boards each time they are moved for upward progress, as the cement or lime indurates them, effectually preventing change of form, and allowing the walls to be built as free from any evidence of distortion, as if purpose-made panels were employed. Potter's, Corpe's, Broughton's, and Macleod's are constructed with some modifications to employ rough boards as described.

An error in judgment is committed by patentees and vendors of concrete appliances by stating that they (the appliances) require but little attention when in use, and that a labourer and boy are able to do all that is necessary thereto ; but in reality it is on the precision with which any appliances are fixed that the accuracy of the entire construction depends ; and patentees know this sufficiently well to employ only skilled mechanics for the purpose when they themselves are erecting concrete buildings.

The necessary qualifications for concrete building appliances may be summed up as these :—

1. They should not be unnecessarily clumsy or heavy, having regard to strength and durability, for

it is not usual that various buildings erected by one person with concrete machines are concentrated in a very limited area ; generally, it is the opposite of this, and points to the necessity of economizing, where practicable, both the bulk and weight of all description of building plant and machinery. When sufficient patented appliances for a pair of labourer's cottages weigh nearly two tons (as do some of those mentioned) it is evident that the carriage of same must form a heavy charge on the builder ; this, therefore, tends to show that if common boards could be employed instead of frames or panels, and which moreover could be made available at the completion of the work for some general purpose, the saving in weight must be very considerable, and that if the standards are of iron and designed with a view to strength and lightness, the bulk must be only a fraction of that of appliances constructed on the original patented system. The actual weight of patented appliances adapted for using rough boards average seven pounds to every lineal foot of walling requiring to be built at the same time.

2. They must be free from all complicated portions. Concrete apparatus made as a model may satisfactorily answer every possible requirement, but it is quite a different matter when exposed to vicissitudes of weather, the rough treatment sure to be experienced where the lowest class of workmen are employed, and the strains and violent usage consequent on the peculiar nature of the work.

3. Concrete appliances should be composed of as small a number of parts as possible, and, as far as practicable, interchangeable—not formed of an almost

indefinite number of wedges, stays, bolts and nuts, screws, pins, &c. and each of different dimensions. Appliances should also be avoided that necessitate large apertures in the walls at every few feet for bolt cores or similar objects, as they tend to weaken them, require no inconsiderable amount of cement and labour to fill the cavities, and are often the means of permitting drifting rains to find a way to the inside of buildings. The use of screws and screw bolts and nuts should be discarded as much as possible, for the effects of the wet concrete and exposure to the weather cause them to rust and become "set," and thereby considerably enhance the time occupied in an ordinary way for fixing and removing the concrete appliances.

The cost of patented apparatus varies from 6s. to 12s. per lineal foot, measuring the actual walls it is capable of forming at one time.

H

CHAPTER VII.

ALTHOUGH concrete made in the manner previously described is probably as economical in point of labour as if mixed by machinery, yet it may occur that manual labour is undesirable. But, as a rule, the walls of ordinary buildings cannot be constructed without interruption, as there are the movable frames or appliances to readjust, window and door frames to fix, and floors and roofs to provide for. Where, however, the work is continuous, as in retaining or dock walls, heavy foundations, and similar constructions, concrete-mixers are advantageous. Messent's patent consists of a vessel, or mixer, of cast-iron, working on a shaft, a water-tank fixed over and near to it, and a hopper attached to a swinging jib, the whole being connected to a stout wood frame, moving on ordinary or on tramway wheels. In the latter case, trucks to carry the aggregate usually run on the same tram-metals as the mixing-machine; the materials for one charge are filled into the hopper, which is turned over the top and discharged into the mixing-vessel, and into which the contents of the water-tank are also emptied. The mixing-vessel is then set in motion, and in seven or eight revolutions the materials are amalgamated, the door at the bottom is opened, and the vessel is emptied almost instantaneously. This machine was invented to make concrete for the blocks used in the Tyne Pier, Tynemouth, and costs from £125 to £135, according to requirements.

Ridley's concrete-mixer is a cast-iron cylinder, keyed on to a central shaft, and mounted upon a strong wood or iron frame, with or without travelling-wheels. The cylinder is placed at any desired inclination, according to the work to be done; the upper end is partially open to admit the materials, which may be fed in continuously, the interior being fitted with shelves running parallel with the central shaft and connected thereto, and which, as it revolves, is, by means of the shelves, continually lifting the ingredients and as continually dropping them. The lower lid of the cylinder is open, and has a spout under it upon which the concrete flows in a continuous stream, and may thence be conducted to where required.

With wheels and axles for travelling, the cost is about £100.

Stoney's mixer is an open trough of wrought or cast-iron. The lower portion is semi-circular in cross-section, and the sides above are slightly splayed outwards. Through the centre of the trough is a wrought-iron shaft, in which adjustable blades are inserted, the blades being so arranged that they may have a tendency to screw the concrete forward as the shaft revolves.

The travelling movement is accelerated or dimi-

nished, as may be necessary, by inclining the trough
more or less towards its delivery end.

Hand-power cannot be applied with advantage, but
a belt from an engine, which may be also employed
pumping water or crushing aggregate, is necessary—
or a three horse-power engine is sufficient, if required
for nothing else.　This machine was invented by Mr.
B. B. Stoney, for special use in making large concrete
blocks for the harbour works in Dublin.

STONEY'S CONCRETE MIXER.

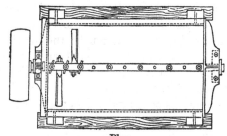

Plan.

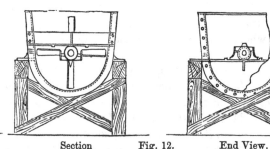

Section　　　　Fig. 12.　　　　End View.

It will be seen that the three machines described,
act each in a different manner to attain the same
result.　The first violently agitates the materials,

simply from the effects of the mixer itself being made to revolve on its own axis; the second lifts a portion, turns it over, and allows it to drop, when it is immediately taken up again, and the process repeated; and the third stirs the materials, passing them on to the mouth of the machine at the same time and in the same manner as a mortar-mixer, and without any violence. For concrete buildings the latter machine would probably be more economical, effective, and convenient, than the two first described; and where steam-power was required for other purposes as well—the size of the buildings to be erected would justify its use—time was an important consideration, or labour at all scarce, an inexpensive and effective concrete-mixer, like " Stoney's," is advisable. And as it answers also as a mortar-mixer, or maker, it serves a double purpose.

CHAPTER VIII.

FLOOR CONSTRUCTION.

THERE is no material that can fairly compare with Portland cement as a matrix for floor construction, where cost, strength, and fire-resisting properties are conjointly the points to be determined, and whatever objections may be urged against Portland cement concrete as a wall material, they have no *locus standi* where floors are the debatable point. There can be but little doubt that, eventually, the essential requirements for a material that shall replace the primitive wooden floor, which as soon as constructed is subject to incipient decay, creates an uninterrupted run and retreat for many descriptions of vermin and insects, and invites and feeds the flames that are continually levelling buildings to the ground, will be found to be possessed by Portland cement concrete. Unlike brickwork, it exercises but little thrust on the sustaining or abutment walls; for moderate spans, it can be applied with so little rise, or arch form, that it has then virtually none whatever, and in many instances it could be applied in conjunction with a judicious use of iron, at the same cost as a wood floor, but to possess treble the strength, and unlimited durability. Concrete does not appear to have been employed for any description of partial self-sustaining floor or ceiling until quite modern times. The first patent in that direction was obtained by James Frost, in 1822, and is called, a "new method of casting or constructing foundations, walls, ceilings, &c.," and the mode of procedure, as far as regards floors or ceilings are concerned, is thus described :—

" Ceilings, whether arched or otherwise, are divided
" into compartments by iron ribs, these being fur-
" nished at the lower parts with small rims or mould-
" ings to receive and support such compartments, and
" the moulds must be surcased inside before use with
" some substance which will prevent adhesion thereto
" of the materials employed for the work, the moulds
" being of course removed as the materials harden.

" The moulds are fixed by means of supports or
" braces in the situation of the structure required
" so as to bound exactly the surface of the intended
" work. The complete structure may then be formed
" by using a quick-setting calcareous cement, having
" hard and durable substances embedded therein, as,
" for example, Roman cement mixed with bricks, tiles,
" stones, gravel, or shingle."

After that date nothing of any importance relative to concrete or beton floors, considered of sufficient public interest to obtain a patent, occurred till Mr. Dennett's " improved method of fireproof construction rfo buildings by the use of arches formed of concrete " was patented in 1863, and described as follows :—

" This invention consists in the adoption of arches
" made or formed of concrete to the construction of

" buildings or parts of buildings ; the object being to
" render such buildings fireproof, combined with great
" strength, durability, and economy in cost of con-
" struction. The concrete of which these arches are
" made is composed of sulphate or carbonate of lime
" together with broken calcined cinders, bricks, or
" other porous material, in such proportions as may
" be found suitable for the particular purpose to which
" the arch may be applied, and laminated with either
" wood or iron. The precise proportions cannot be
" given, as they will differ according to the greater or
" less strength required to be given to the arch."

The remainder of the specification describes the
buildings for which concrete floors, ceilings, &c., are
most suitable.

A Mr. Turner in 1863 also patented " an improved
" fireproof floor or roof for buildings, bridges, and
" other structures," and in which the embedding of the
iron girders in ordinary concrete, so as to completely
envelop them, appears as the principal object of the
invention ; not as in Dennett's, where the material
composing the concrete was the subject matter of the
patent.

Within the last ten years patents have been granted
for fireproof concrete floors to Homan, Phillips, Fox
and Barrett, Brannon, and several others, and where
the general principle appears to be the economic distribution, in various forms, of wrought iron, principally as girders or beams for supporting the concrete
at intervals, and in which the said girders or beams
are, according to requirements, either placed beneath,
partially embedded therein, or wholly covered and
encased by ordinary concrete. In the latter instance

only can concrete floors be said to be fireproof, for it
is a well-known fact that the change of form iron is
subject to when exposed to extremes of temperature,
renders it one of the most treacherous materials
employed in buildings—should a fire occur—unless
encased with some non-conducting material like bricks
or concrete. Brannon's system consists of forming a
skeleton or framework of wire netting attached to iron
frames, and the whole embedded in the concrete. The
method, as described by the inventor, is briefly this :
" My invention has reference to the mode of forming
" roofs, floors, ceilings, stairs, doors, walls, and other
" parts of buildings or other structures of cement or
" concreted materials in combination with metallic,
" fibrous, or laminated substances, with a view to
" render them more durable, fireproof, and healthy,
" and consists in employing for the said purposes a
" sustaining metallic framework or skeleton firmly
" fixed and bolted or bound together, and upon which
" is stretched wirework, so as to partially enclose or be
" completely embedded in the said concreted materials
" which compose the body of a structure or any part
" thereof, and thereby perfectly bonding the same into
" a solid and coherent mass ; and in order to still
" further increase the coherency of the masses, and
" render the same capable of resisting transverse
" and tensile strains, intimately commingle with the
" concreted materials, and interpose between layers
" thereof or between layers of bricks, tiles, slates, or
" stones—wire, or fibrous substances."

A simple way of employing concrete as—to a certain
extent—a fireproof material is shown by Fig. 13,
representing a section of an ordinary floor. A A are

common iron clips, about $1'' \times \frac{3}{8}''$, which are let into
the top edges of the joists so as not to project above
them and interfere with the flooring to be afterwards
laid. These can be placed about every three feet apart.
Common roofing tiles, B B, cut to the same width the
joists are in thickness, are nailed over the whole under
edges, but having had a bed or layer of mortar inter-
posed between the two when the tiles are about to be
nailed on. Rough boards are then placed temporarily
against the tiles to support the concrete, which is
deposited thereon about two or two and a half inches
in thickness. When the concrete is set sufficient to
allow of the boards being removed they are refixed in

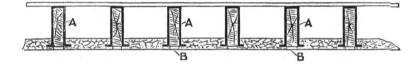

Fig. 13.   Scale $\frac{1}{2}$ inch = 1 foot.

another place. This work could be entirely executed
by an intelligent labourer, the cost being, under any
circumstances, not more than about 3s. per superficial
yard. The extra weight on the floor would be about
18 cwt. per square of a hundred superficial feet, and
for which a slight increase in the dimensions of the
floor joists would compensate, the cost of same being
more than counterbalanced by no lathing being necessary for ceilings, and no sound boarding for floors.
The advantages gained by treating floors thus are that
it renders the under sides, when plastered, practically
fireproof, lathed ceilings are avoided, and cracks and
disfigurements so often caused thereby provided

against. Moreover, if about six inches of sawdust is deposited on the concrete the whole floor is, in a great measure, impermeable to sound. For an aggregate for this purpose nothing can be more suitable than coke from gas works, or mill cinders.

There are numerous methods of constructing iron and concrete floors in a simple and economical manner, the best being by using ordinary rolled T iron girders, and depositing concrete thereon, either flat, or arch-formed on the under side, and employing rough boards as a means of temporarily sustaining the concrete until it has sufficiently hardened to allow them to be removed. Where the concrete is formed flat, or of equal thickness throughout, the bearings or supports must be closer than if arch-formed, as necessarily the distribution of the concrete is not of that form that tends to give the maximum amount of strength with a minimum amount of materials, and should only be practised when some reasonable objection to an arch or curve-shaped under-side of floor exists. None the less a flat or slab concrete floor may be very readily made, and possesses considerable strength, as the following experiment will testify :—

A slab of concrete measuring 6 feet in length by 4¾ feet in width, equal to 28½ superficial feet, by 5 inches in thickness, and composed of 2 bushels of Portland cement and 9 bushels of crushed slag (the refuse from iron ore), was placed on supports, whereby the two long sides of the slabs had each solid bearings of two inches, and the other two sides were free. One month after being made the slab was loaded with 550 bricks, together weighing 3,725 pounds, and also subjected to considerable impingement, but which conjointly failed

to break it. No other means being at
hand to increase the strain, and the
slab requiring to be removed, it was
ultimately broken to pieces with sledge
hammers. The weight of the bricks
alone was equal to 140 pounds per
superficial foot, clear of supports, and
as the strength of the slab would ultimately have been at least double that at
only a month old, its carrying powers
would have been, probably, a safe load
of 270 or 280 pounds per superficial
foot. Fig. 14 is a part longitudinal
section of a floor over stabling, and
which is 96 feet in length and 17 feet
in width, clear of external walls.
The mode of construction and details
of cost are given as an illustration of
the economic use of iron and concrete
for common floors suitable for various
buildings. Four division walls assist
in supporting the floor, and necessarily reduce the number of rolled
iron joists required. By employing
nine of the latter (each spanning the
width of the building, 17 feet, and
having a bearing on the walls of 6
inches), the distance from centre to
centre of supports averages about 6 ft.
9 ins. The iron joists are 9½ inches
in depth, the top and bottom flanges
3¾ inches in width, the thickness of
the webs ⅜ inch, and the weight 25

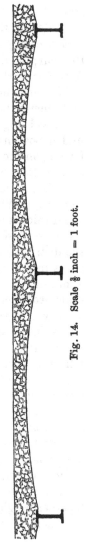

Fig. 14.   Scale ⅜ inch = 1 foot.

pounds per lineal foot. The concrete, as shown by
section, averages about 5 inches in thickness,
weighs 1 hundredweight per cubic foot, and is composed of 6 parts by measure of old bricks crushed to
pass through a 1½-inch mesh sieve, and 1 part of the
best Portland cement. Each joist is capable of sus-
taing a safe distributed load of 4¾ tons, and previous to
being fixed in place receives two coats of oxide of iron
paint. The total cost of a cubic yard of concrete for
this special purpose is obtained as follows :—

	s.	d.
Crushing aggregate by steam power		8
Four bushels of Portland cement, allowing for shrinkage @ 2s. 9d.	11	0
Labour of mixing and depositing in place ...	2	0
Use and waste of materials in timber supports, and labour in preparing, fixing, and remov- ing same	3	0
Finishing surfaces of floor and soffit, as de- scribed hereafter	3	0
	19	8

In the case in point, rough materials for aggregate are on the spot, and to be had for nothing—in
fact, would otherwise have to be removed. This
advantage, however, has but little effect on the total
cost of floors generally, the bulk being so small compared with that of walls, for a cubic yard of old
materials would suffice for six superficial yards of
flooring similar to that shown on section

The rough surface of floors is treated thus : Neat
cement mixed with water to make a stiff mortar is applied with a plastering trowel, and a hand-float, or piece
of thin wood, about 9 × 4 inches, having a handle,

is then used to thoroughly work the cement into all the crevices of the concrete, but leaving *no body or coat of cement* on the surface; the steel plastering trowel is then used to give it a smooth surface, and the operation is finished. The under side of floor is treated in a similar manner, except that the cement mortar is composed of three parts of clean sharp sand and one part of cement, and not trowelled to a smooth surface. The concrete must have become quite hard previous to this treatment, or some portions may be loosened. The effects of this manipulation are, 1st, no thin coat of cement is left to eventually peel off, but it forms an integral part of the concrete, and creates an almost unwearable surface; 2nd, ordinary labourers can readily learn and easily perform this kind of work; 3rd, economy in cost, which amounts to about 4d. per superficial yard. Obviously the floor has not so even a surface as if done by the ordinary method of hand floating and trowelling by a competent plasterer, but it has sufficient for the floors of granaries, barns, warehouses, maltings, &c.; and in point of durability is superior to a coating of cement and sand of the usual thickness.

Assuming, for sake of example, that the concrete covers the entire area of the floor (but which, however, it does not do in the case in point), no abstruse calculations, with the details of construction already given, are necessary to arrive at the cost of such flooring complete, and the load it will carry. The concrete being estimated capable of sustaining a greater weight with safety than the iron joists, need not enter into this calculation. The latter cost £12 per ton delivered on site. The actual outlay, based on these figures, would.

amount to £47 14s. The area covered being 1,632 superficial feet, would give the prime cost per square as £2 18s. 6d., or 5s. 3d. per superficial yard; and after deducting the weight of concrete, the safe live load could be calculated at 32 tons, or nearly 2 tons per square. Comparing the cost with that of an ordinary wood floor, the following would be a fair estimate for the latter :—

```
No.   '   "     "     "       '    "
  6   18 0 × 9 × 6 =   40   6 binders.
 15   98 0 × 8 × 2 = 164   7 joists.
  6   18 0 × 4 × 3 =    9   0 wall plates.
                     ───                  £    s.   d.
                     214   1 cub. ft. @ 2/6 =  26  15   3
        '   "    '    "
      96 0 × 17 0 = 16¼ squares, 1¼ groved
          flooring @ 27/0    ...     ..    = 21  18   9
187 lineal feet herring-bone strutting @ 3d. =  2   6   9
                                               ─────────
                                               £51   0   9
```

So that a common deal floor on ordinary Baltic fir timbers would cost more than one of iron and concrete. But it may occur that for some purposes a common description of wood floor is desirable, and an inexpensive way of effecting this is to use small blocks of deal about 6 inches in length by 2 inches in width, and 1 inch in thickness; and instead of finishing the floor as just now described, these blocks should be bedded with Portland cement on the rough surface of the concrete, and the crevices between them "grouted" with plaster-of-Paris. The points to be observed are that the blocks should be laid in the manner known to paviors as "herring-bone," so that the expansion

caused by the moisture contained in the plaster
is distributed in equal directions; also that the
" grout " over and above that which is necessary for
filling the crevices should be as quickly as possible
absorbed by dusting the surface with dry sawdust.
Of course, any kind of wood will serve as well as deal,
care, however, being taken to avoid any with large
knots, or that has a natural tendency to warp, as elm,
&c. The additional cost of this would be about 2s.
per superficial yard. Fig. 15 is a section of another
description of floor, in which iron, concrete, and wood
are employed. In this example the iron joists are
entirely hid from view, being taken up partly by the
concrete and partly by the wood joists. The latter
have a bearing on the concrete, and therefore need be
only of small dimensions. For ordinary house floors
the span may be 8 feet to 9 feet as shown, with an average thickness of concrete of 5 inches; this span would
for ordinary rooms make one iron joist only necessary.
If the concrete is formed as a flat slab, the span should
not exceed 7 feet, and 4 inches in thickness would
be sufficient. The wood joists should be notched to
clip the top flanges of the iron joists, and thereby
become rigidly fixed; and if sawdust be deposited on
top of the concrete to the under side of the floor boards,
a fairly sound proof floor is obtained.

Floors where the iron joists are embedded in the
concrete are capable of sustaining heavier loads
than when merely supported on the top flanges of
the same; for, if not held rigidly in position through
their entire length, joists will " buckle " or twist
under heavy pressure, previous to any sensible deflection taking place. Fig. 16 shows a form of construc-

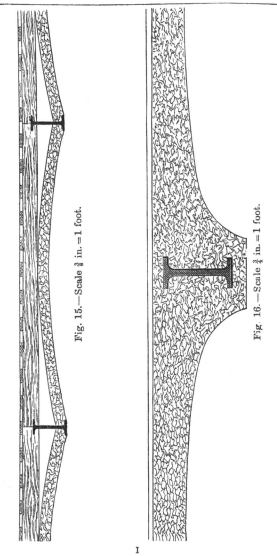

Fig. 15.—Scale $\frac{3}{8}$ in. = 1 foot.

Fig. 16.—Scale $\frac{3}{4}$ in. = 1 foot.

I

tion where the iron girder being deeper than needed
for the floor, the latter is fairly rendered fireproof by
causing the concrete to envelop the unprotected portion of the iron. The concrete is sufficiently coherent
in itself to clothe the exposed part of the girder without danger of coming away in fragments. The sketch
shows that almost any form of soffit could thus be
obtained at small cost, and that would admit of effective treatment in many ways. It will be apparent to
everyone that the depth of iron girders must, in the
majority of cases, be of necessity greater than the
required thickness of concrete, so that it becomes
necessary to treat them—if protection from the effects
of fire is required—in one or other of the ways
described, and even then there is the element of danger existing, which, by some unforeseen circumstance,
may become the indirect cause of greater loss than
even a fire itself might create.

An incombustible floor is not necessarily a fireproof
floor, and we have yet to learn the action of intense
heat on concrete encasing iron, where large surfaces
and heavy masses are the rule, and where it may be
exposed to the sudden and intermittent application of
immense volumes of water from powerful fire-engines.
The failure of a few cubic feet of concrete so as to
expose only a small portion of naked iron may be the
precursor—the thin end of the wedge—that may
eventually cause the destruction of whole blocks of
buildings. This may be looking at the matter from
an extreme point of view, but it is well to do so when
the importance of the subject is considered.

It becomes a question, therefore, in floor construction whether iron might not be dispensed with alto-

gether, or at least in many instances where moderate
spans from wall to wall occur, and also whether the
value of the iron thus saved might not with advantage
be employed to increase the quality and consequent
strength of the concrete itself.

Fig. 17 is a plan and section of a trial floor of concrete where no iron was introduced, the materials composing it being six parts of broken bricks and one part
of the best Portland cement. It will be seen that the
thickness of the concrete at the haunches is only 7
inches and at the crown 5 inches, the sustaining
walls being 9 inches in thickness.

SECTION.

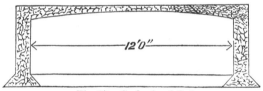

Fig. 17.—Scale $\frac{3}{16}$ = 1 foot.

PLAN.

It is not always practicable to give floors the advantage of direct support on all four sides—stairs, windows, &c., may intervene and leave a portion of one side to carry itself. To allow for this probability, three walls only were built to bear the floor. The trial was made at a month old, the result being that with a load of 10,000 pounds equally distributed, but clear of the sustaining walls, and some considerable impingement practised, the concrete remained sound, but with 14,000 pounds and no impingement the floor broke, the rupture occurring where shown on section by a dark line.

This conclusively proves that, except to carry very heavy loads, iron could be dispensed with in the majority of cases, for spans up to say 16 feet at least might be safely covered, and if formed in the shape of a vault or groined arch, where practicable, there would be scarcely a limit to the void that might be covered.

The various forms of floor construction that have been described are the most economical of their respective kinds, and will compare favourably for strength, cost, and simplicity with any others, whether patented or otherwise. The French adopt for their beton floors a plan of combining small T-iron bearers crossing the principal iron joists or girders, and resting on their bottom flanges, or otherwise connected thereto, so that they become embedded in the beton, or concrete. These are, perhaps, absolutely necessary with the matrix employed —plaster of Paris—which can be cheaply procured; while, on the other hand, Portland cement is more costly than in this country.

An error often committed—for, however, the very cogent reason for protection in case of fire—is to have the concrete of ordinary segmental arches, of sufficient thickness at the springing or junction with the iron girders to completely envelop the latter; but the immense weight thus created and thrown upon the sustaining walls for no ulterior purpose whatever but the one named, and the enhanced cost, render this plan commendable only when a positively fireproof structure, as far as circumstances will permit, is imperative, and where cost also is a secondary consideration.

There is scarcely a doubt but that concrete in itself is capable of withstanding the attack of fire as well, or better, than any other material applicable to wall building and floor construction, but there has been no instance on record, at least publicly known, where a building entirely erected of concrete, and having concrete floors—no naked iron being exposed—has been subjected to a conflagration similar to those that occasionally throw a glare over half the city of London. The trials on a small scale, simply by heating small masses of concrete, or by burning a few cartloads of inflammable materials in proximity to a concrete wall, cannot be indicative of the behaviour of the same material when tied up in form and shape in broad masses, and exposed alternately to fire and water. And it must be remembered there is concrete, and concrete. Flint gravel, or crushed chalk flints, if exposed to fire, will either calcine or split up into fragments; nor will the admixture of Portland cement prevent this, even under a moderate continuous application of heat. Where smoke flues are formed with cores (subsequently withdrawn), and the concrete is composed of either of

the materials just named, chimneys possessing a
number of flues with but slight divisions between, will,
before many years, possess one only, of an approximate
size to those constructed by our forefathers. Chimney-
sweeps accustomed to concrete chimneys readily learn
of what materials they are made from the *débris* that
reaches the bottom after being swept ; and they state
that where flint or flint gravel concrete is used, one to
two quarts from each flue, and each time they are swept,
is not at all an unusual circumstance, but that where
broken bricks, mill cinders, slag, or similar ingredients
form the aggregate, little or no dislodgment takes place.

For fireproof floors, chimneys, &c., no doubt the
best possible aggregate is crushed fire bricks, or Staf-
fordshire fire clay, burnt hard and broken small ; then,
in order of merit, probably come crushed hard-burnt
stock bricks or clinkers, mill cinders, coke or slag,
and broken ordinary bricks. Portland cement loses
but little of its capability for withstanding tensile
strains if heated to a red heat, and then allowed to
cool gradually, but it becomes brittle ; there is no
doubt, therefore, but that cement concrete in floors,
after exposure to intense heat, would be less capable
of withstanding sudden strains or violent con-
cussions. Blocks of concrete composed of broken
bricks and Portland cement, after several hours' incar-
ceration in the furnace of a steam-engine, retained
their property of cohesion, apparently without any
diminution.

Sulphate of lime or plaster-of-Paris has not been
credited as a material that would resist the vicissitudes attending a large building when on fire, or
retain the properties necessary for a matrix in concrete,

under similar conditions ; and Dennett's patent material, which is largely composed of sulphate of lime,
although advocated as possessing all the needful qualifications, has apparently not been under trial to any
great extent. In the *Builder* for Dec. 28th, 1872,
a correspondent says that, "Although the Dennett
form of construction is a clever adaptation of a cheap
and useful material, it will not resist fire, for the effect
of water being then poured upon it would be to reduce
it to a mass of rottenness." This opinion is corrobo-
rated in a later number of the same journal, by a
" Plasterer of thirty years' standing ;" and as so direct
an assertion in a high-class journal remains uncontra-
dicted, we may assume there is some truth in it.

In forming the supports for concrete floors nothing
more is wanted than some rough boards of regular
thickness, properly upheld, and not liable to "swag"
with the deposition of the concrete. Fig. 18 shows
an inexpensive way of doing this, by making the bottom flanges of the iron joists carry the supports. Of
course this cannot be done where the concrete is disposed as in Fig. 15, but must then be by upright
posts or stanchions, resting on blocks or cills laid on
the ground, or floor beneath.

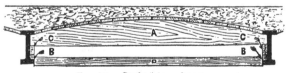

Fig. 18.—Scale ⅜ in. = 1 foot.

A is the centre or rib of sufficient strength to carry
the concrete, and also withstand gentle impingement
of same with little or no vibration ; B B are pieces of

timber or quarterings resting on the bottom flanges of girders ; c c are folding wedges, the release of which liberate also the ribs and boards ; and D is a stretcher of any slight scantling, which, when knocked away, allows the timbers, B B, to be removed, and the whole to be refixed elsewhere. For spans of about 7 feet ribs may be cut from 8 in. × 3 in. boards, and placed about every 4 feet apart, and the "laggings," or cover boards, should be one and three quarter inches thick : these dimensions are for concrete averaging about 6 inches in thickness ; any increase in the amount of concrete necessitates a proportionate increase in the dimensions of the temporary timbers. The boards on which the concrete is to be placed are better for being planed, but this is not an absolute necessity, and they do not adhere so much to the concrete if brushed over just prior to use with a solution of soap-and-water of the consistency of cream. The boards, if dry, should be laid with a space between them of a quarter of an inch for every 7 inches of wood, to allow for expansion, and also to provide means for the discharged water to escape, and permit of more easy removal than if they wedged each other together.

Some effective forms of finish to the soffits of floors may be practised by lying on the boards encaustic tiles in patterns, plain glazed tiles, or some similar materials ; this, however, necessitates absolute truth and rigidity in the centreing, or temporary bearers and boarding, as there is no possibility of afterwards altering the position of any of the embedded substances. If the boarding is uneven in thickness, an uniform surface for the tiles to rest upon may be made by depositing on the boards a layer of sand, or a bed of weak mortar, and on which they can be laid in correct position.

Should the cement ooze throught the joints of the tiles and disfigure their surface, it cannot be removed till the boards are taken away, by which time it will have got hard, but such portions as then cannot be scraped off without injuring the tiles may be entirely displaced by an application of diluted muriatic acid. For small or intricate designs, where the colours of the tiles are not distinguishable on the under sides, the latter should be coloured with a colour wash resembling the face sides, so as to facilitate the process of laying them, and ensure correct execution of the design, and a little thin cement poured over previous to depositing the concrete insures them from being disturbed thereby.

For a plastered ceiling, concrete offers an excellent key for ordinary plaster, or for Keene's or Parian cement for painting. It will be seen that a concrete floor provides facilities for the execution of almost any design or description of plain or ornamental ceiling, with much less trouble and considerably less cost than by any other known method, and at once disposes of the question of " æsthetics " as far as concrete ceilings or soffits of arches are concerned.

The preparation of concrete for floors differs in no way from that practised for walls ; the proportions of materials must depend on the judgment of those superintending the work, and on the quality and description of the aggregate, but as a rule six parts of the latter and one of best Portland cement give excellent results.

The materials should be subjected to a gentle beating or impingement with a common turf or grass beater, having a piece of zinc or sheet iron fastened on the face to prevent adhesion, and this impingement should be continuous as the concrete is being deposited, that it may all undergo equal treatment. If the floors are formed previous to the walls being built above them, so that in fact the walls recommence on the floors—and which practice gives by far the best results, as no corbelling, indents in walls, or other forms of support for the wall ends of floors are required—care must be taken that the latter be either covered till the cement is thoroughly hardened, or that they be grouted with liquid cement, otherwise a sudden storm may wash and carry a considerable portion of the matrix through the floor before it has had time to set, and irreparably destroy the quality of the concrete. No temporary supports for concrete floors similar to the illustrations should be removed under a week in summer and a fortnight in winter; but this must depend considerably on the nature of the cement, whether slow or fast setting, and on the description of aggregate, &c., and the condition of the surfaces of the concrete exposed to the drying action of the sun and wind must not be accepted as a criterion of the condition of the bulk of concrete generally.

It is a common practice to " pack " concrete arches. This consists of introducing boulders, brickbats, or large-sized fragments of any similar materials into the concrete mass. The contractors' theory for this process is that it creates a bond to the bulk of the work, but the real motive more often than otherwise is that

it effects a considerable saving in the amount of concrete used, and at a nominal cost. This tendency for
economizing the more costly material often leads to
an immoderate use of packing, and although a limited
amount would do no harm it would do no good, and
therefore the safer plan, especially when the arches
do not average more than 6 inches or 7 inches in thickness, is to entirely prohibit the practice. In depositing the concrete in place it is desirable to finish one
arch, or one portion between supports, with as little
delay as possible, as the weakest part of a concrete
mass, whether wall or floor, is where the incoherent
material is joined to another portion, wholly or partially consolidated; in other words, the concrete should
not be deposited in layers or fragments, but be as far
as practicable disposed *en masse.*

The slight change in form Portland cement concrete
is subject to in setting, is insufficient to have any prejudicial effect upon the external walls unless the
floors are of very large dimensions, and then, probably, one portion has consolidated and the change
taken place before another portion is commenced. It
is an essential condition that what are known as
" slack blocks " or " slack wedges," that is, ordinary
wedges laid in pairs, with the thick end of one resting
on the thin end of the corresponding one, be employed
in all cases for the purpose of preventing any unnecessary violent treatment in removing the temporary
supports, and that the whole of the latter may be withdrawn gradually and easily. Where these simple but
necessary precautions for floor construction are not
observed, failure must be the result, and a lamentable
case of the kind occurred at Portsmouth in the early

part of 1876, when a floor, 26ft. by 22ft., divided by two rolled iron girders into three bays, fell and crushed four men who were beneath. According to the evidence of the surveyor appointed to trace the cause of this " accident," it would appear that the girders were twelve inches deep, and capable of sustaining an unusually heavy load, the width between the bearings was 8 feet 10 inches, and the bearing on the girders 2½ inches, the concrete averaging 12 inches thick. The concrete was composed of one part of good Portland cement to five of gravel and sand, and was disposed in two layers, *between which* a course of bricks was inserted or imbedded. The centreing or supporting timbers were fixed *without folding wedges*, so that considerable force and concussion was necessary for their removal, and they were in the act of being displaced, *four days* after the work was executed, when the accident occurred ; and it would also appear that *no proper method* of correctly apportioning the aggregate and matrix was adopted. Here, then, was a combination of circumstances that positively invited an accident, if it could be so called, and for which the materials were in no way blameable.

It is not probable that for ordinary floors, patented materials or patented processes are likely to come into very general use, on account of the additional cost and the special means of construction hat must be adopted, but the application of common labour in a common way, as has been here described, is equally as simple and as easily performed as the every-day description of wood flooring, with which everyone is acquainted.

CHAPTER IX.

THE DISADVANTAGES OF CONCRETE.

A DISADVANTAGE that concrete, like all other building materials, is subject to, is expansion and contraction through change of temperature; and although with walls of brick or stone, having mortar joints at every few inches, this change of form may be so slight and so equally distributed as not to be perceptible, there is no doubt that the change exists. In some long boundary walls of brickwork occasionally a slight vertical fissure may be noticed, which is universally ascribed to a " settlement," whereas it is probable that many of these are due to contraction, the result of one place in the wall being weaker than any other. Concrete walls have been built 70 feet to 100 feet in length, the rough surfaces merely filled up with cement, and have shown no fissure, but, instead, a number of minute cracks almost invisible to the naked eye, and resembling in appearance a spider's web; these are so small that they cannot affect either the strength or weather proof qualities of the material. When walls show a perceptible amount of shrinkage in the form of vertical fissures, these spider-webbed cracks do not exist. Where concrete walls are tied

together by the timbers of a building and weighted
with the roofs and other parts, shrinkage is scarcely.
ever perceptible when the work is properly done; but
if it should occur it is naturally where the walls are
weakest, viz. at the window and door openings.
Occasionally window and door heads have shown
symptoms of cracks and displacement, and which, in
nine cases out of ten, is the result of the way in which
the temporary wooden supports, known in brick buildings as "the centres," are treated: the centres (for
want of a better designation) support the semi-liquid
concrete until it has hardened or consolidated over all
openings in walls, in the same way as they would
serve for supporting brick arches during construction;
these centres are often made of dry wood, perhaps
from boards 10 inches or 11 inches in width.   It is a
property of Portland and other cements that, although
possessing so strong an affinity for water, they will
exude all in excess of that required for hydration;
and therefore, after the centres are necessarily fixed
rigid, and the concrete is deposited thereon, they are
subjected, perhaps for hours, to an application of the
water that the cement is endeavouring to part with.
The result is, the centres perceptibly expand, and being
unable to move downwards, rise sufficient to create a
gradual displacement of the concrete during the process of setting, and at the very points where greater
strength is required.   The remedy is obvious: the
centres should be kept in water for some hours previous to use, and spaces left between the boarding of
same, and on which the concrete is deposited, to allow
room for the discharged water to escape.   In every
case this will be found quite clear, not, as has been

supposed by many, that if room be left for water to pass off, it will carry with it some of the cement employed as a matrix.

But there is no doubt that a course of iron hoop, built in all concrete walls, every 2 feet or less in height, as the work progresses, assists in keeping them from giving evidence of contraction, and under any circumstances the cost is so trifling that it should be practised in all cases. Probably the hoops known as "Tyerman's patent" would key into the concrete better than ordinary hoops. With hoop bonds employed for brick buildings it is sometimes usual to specify they shall be tarred and sanded, to avoid corrosion; but this is quite unnecessary in concrete buildings, as no oxidation takes place; bond taken from concrete walls, that had been erected more than a year, was found to be as free from rust as the day it was manufactured.

Extremes of temperature during construction also help to a certain extent to cause shrinkage, although heat is more prejudicial than cold, as the matrix then sets too quickly. When this takes place it is advisable to keep concrete walls occasionally watered, after the lime or cement has set sufficient to avoid any possibility of disturbing it by so doing.

With some fissures it has been found that they do not exist through the entire thickness of the walls; at other times they are much smaller, or disappear entirely in the centre, resembling the "heart shakes" that are often seen on the cross section of a tree or balk of timber. Concrete laid as paving will often show fissures or cracks on the surface, but they seldom penetrate far, as the more equal temperature of the soil beneath prevents this ; and for the same reason underground tanks, vaults, and walls constructed below the surface are generally free from any change of form.

Although protracted delay in erecting concrete buildings is not to be recommended, neither is undue haste ; the "merit" claimed for some appliances that they enable 5 feet or more in height of walling to be erected in one day, induces that high-pressure speed which is as fatal to concrete building as to any other kind of building. The materials also should be of average uniform quality, and when work is suspended for the day, the walls should be left with all portions as far as practicable level, and "jumps" or uneven cavities and ridges in the work avoided. When the walls dry quickly, it is well to water them previous to depositing more concrete thereon, especially if work has been suspended, though only a day. Ordinary care in construction in the manner described will show as a result very little, if any, after evidence of contraction or expansion.

Concrete walls are more liable to be the medium for condensation of dampness in the atmosphere than brick ones, as the surfaces of the former, though dryer, are colder, and being non-absorbent, the condensed moisture is apt to remain on their surfaces or trickle to the floor. To avoid this as much as possible, doors and windows should be closed in damp, and be opened in dry weather, or other means of ventilation secured. A small fire in a room will generally prevent all condensation, and so will a single candle kept burning in a cottage bedroom during the night. Dr. Richardson considers this readiness to

condense moisture an advantage, as it gives evidence
of the atmosphere in the room being damp, and
thereby warns its occupants; moreover, if the walls
abstract the moisture, the air from which it is obtained
must be the dryer. With painted, or papered and
varnished walls, if covered with moisture, a fitting
opportunity offers to thoroughly cleanse the same
with a sponge, and which can be executed without
soiling either furniture or upholstery, and the amount
of "matter in the wrong place" (but which nature
thus provides simple means of removing) that may be
in this way occasionally taken from the walls of a
room, even but little used, and of a country house
apparently away from all sources of dirt, is something
surprising. Where walls are to be papered, a stout
lining paper first attached would tend to prevent condensation, and flock papers more so than those with
plain surfaces.

Possibly the only disadvantage of any real importance that concrete possseses,—and a difficulty that
would appear not easily to be surmounted—is its
capability of conveying sound, and the harder the
aggregate, and stronger the cement, the better conductor the concrete becomes. It is somewhat singular
that Mr. Reid, in his treatise on "Concrete," in enumerating its advantages, should have claimed superiority for this material because of its "impermeability
to sound," and that Mr. Tall, in a Paper read before
the Architectural Association, should also have declared, after having erected a large number of concrete buildings, that "there was no foundation for such
assertion;" whereas a concrete wall 9 inches thick
really conveys sound about equal to a brick wall of
half that thickness; but lime concrete is not so good a

conductor as cement concrete, and this would seem to point to the principle of employing the former for all internal or partition walls, even if the external are made with the latter. For ordinary houses, a lime concrete wall 9 inches in thickness, would not pass sufficient sound to be considered a serious evil, and the plastering on both sides would also assist in deadening it. It has been suggested to batten and lath-and-plaster concrete partitions, to prevent sound passing through, but the evils attending battened walls do not permit recommending this; probably the better way would be to build them hollow, but this incurs considerable more trouble with concrete than with brick; the only practical way being to insert taper wood cores, one or two inches in thickness, and of convenient size for removal, in the walls during construction, withdrawing them as the work proceeds, or the insertion of sheets of dry hair boiler felt in the middle of the concrete. To stiffen the felt, it might be tacked to thin rough boards, of convenient lengths, both boards and felt being built in. But the disadvantage for dwellings of conveying sound is a matter of no inconsiderable gain in the case of concert-rooms, theatres, lecture halls, and other public buildings; for concrete walls create a reverberation, or sonorousness, superior to any other wall materials.

The claim by opponents to concrete construction, that great difficulty exists in making future necessary alterations, is surely an advantage, as not only does it show that it is more difficult to get to pieces, therefore more substantial, but it is a specific against the mania for perpetually " adding to " and remodelling, which some persons seem to possess, solely for the benefit of the builder.

CHAPTER X.

FOR ordinary buildings, roof construction with concrete is about the least fitting purpose for which it can be adopted, and for several reasons.

The great weight thrown upon the supporting walls, the sightlessness of flat or continuous curved roofs, (for an arched roof of irregular plan would involve no inconsiderable difficulty,) except in town houses where a parapet wall may possibly hide them, and the effects of a variable temperature, from which floors are in a great measure exempt, are all against the material. Still there may be instances where a flat roof is desirable, or even compulsory, as for the erection thereon of an observatory or photographic room. Even a garden or conservatory may be a necessity under certain conditions.

The flat form of roof cannot be constructed better than by the introduction of iron joists, as described for floors, where the intervals between supporting walls are greater than may be considered safe to span with a concrete slab alone.

Fig. 19 is a section of a concrete roof applicable to a row of buildings of uniform construction, where the division walls are equally distant apart, and not more than 18 feet span. A A are temporary pieces of timber about $9'' \times 4''$—built in the walls about every 4 feet apart to carry the centreing and concrete, and B is a chimney flue. A flat roof on the same principle may be constructed by giving the concrete less rise.

This plan is similar to that adopted for the roofs of the houses built on the Redcliffe estate at South Kensington, except that the latter are covered with common tiles, bedded on each other and breaking joint, the surfaces afterwards being cemented. The timbers are also left in, thereby providing for a lath-and-plaster ceiling, but for a concrete roof, space

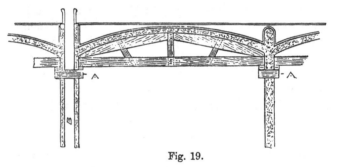

Fig. 19.

would be gained, materials saved, and other advantages accrue, by letting the soffit of the concrete arch form the ceiling.

Roofs constructed of concrete, whatever form they take, must necessarily have temporary wooden supports, the same as for floors, and the general principle as far as regards the nature, quality, and relative proportions of the materials, and method of disposing of same, that is applicable to floors, is also suitable for roof construction.

It is desirable to avoid metal gutters in conjunction with concrete, owing to the expansion and contraction that lead, zinc, and other metals are subject to when exposed to change of temperature, and it is better, therefore, to make such water channels as cannot be avoided, in the concrete itself, but afterwards carefully rendering them with Portland cement and sand; and this should also be done to the whole superficial area of concrete exposed to the weather.

Flat top roofs, owing to the extremes of temperature to which such exposed surfaces are liable, and inability to conform to these changes, but to which a vaulted roof is capable to a great extent of yielding, should be covered with a coat of asphalte, for the comportment of that material under climatic changes is more favourable than concrete.

The avoidance of thrust on the sustaining walls, or, at least, to a very great extent when compared with brick arches, almost unlimited durability, and possessing also the desirable properties of being fireproof, and a non-conductor of heat and cold, are all important points in favour of concrete as a roofing material; but, on the other hand, are the undeniable facts that for ordinary buildings it would be more costly than other forms of roofing (at least, with our present knowledge of dealing with the material), be from five to ten times the weight of common tile coverings, require a substantial wood support of the exact form of the roof and over its entire area, and for temporary purposes only, and be anything but pleasing to the eye.

Under all circumstances, therefore, it is not probable that concete will come yet awhile into general use as a roofing material.

CHAPTER XI.

GENERAL REMARKS ON MONOLITHIC WALL BUILDING.

CONCRETE block building is obviously very similar to ordinary brick or ashlar stone construction, except where patented forms of blocks, as Sellar's, Broom hall Company's, or other known methods are adopted, and in such cases the process is fully described to users by the patentees or their agents; but it is with unpatented processes of monolithic construction that some information is needed.

When temporary wood appliances, or patented apparatus are employed, the first thing necessary is that they should be arranged so as to properly delineate the intended walls. Particular attention should be directed to the standards that they are perpendicular, and that the exact size and position of walls shown on plan are faithfully carried out, nor must the plea of "near enough" be accepted as a sufficient excuse for one wall being too long or another too short, one leaning a trifle out and another in. It becomes therefore necessary to have the plans of all intended buildings drawn carefully to scale, to avoid probability of error, and better still for the fixer to have every dimension and distance figured on the

plans and sections, that he may know where door or
window frames are necessary, chimneys are required,
party walls occur, &c., as it must be remembered that
the rate of progress is much greater with concrete
than with bricks, and that infinitely more trouble
occurs to rectify errors in construction,—two reasons
why every information in connection with the drawings of concrete buildings should be lucidly given. It
also saves time, and facilitates the operations of the
" fixer," if the position of the standards are marked on
the plans previous to commencement. A carpenter
with sufficient knowledge of ordinary drawings to
enable him to know when all parts of a building are
in accordance therewith, is the most fitting man to
superintend this work, and who requires a labourer to
assist him in the more laborious portion of fixing and
removing the appliances. The cost of this part of the
work depends on the nature of the building to be
erected—whether simple on plan or the reverse,
whether of moderate height or of lofty proportion,
and also if the walls are slight or massive ; but the
average cost may be reckoned at between 6d. and 1s.
per cubic yard, exclusive of the value of fixing window
and door frames, and other incidental works that
pertain to buildings, whatever their materials of construction may be.

The class of men most suitable for mixing the
materials and filling in the frames are necessarily
active and able-bodied labourers ; the work of itself
is laborious and monotonous, and is deserving of fair
recompense. The best description of men for the
purpose are—strange to say—not ordinary town
labourers used to builders' work, but, where they can

be obtained, the better class of agricultural labourers, who may, perhaps, have never previously been employed on a building. They will take more interest in their new calling, are, as a rule, expert with the shovel, require less wages, and are more easily trained to the work, having no groove to travel out of, or possessing set notions on matters pertaining to building processes, as one-half of the ordinary builders' labourers in towns are sure to acquire after a short initiation into the mysteries of mortar-making or hod-carrying.

When the concrete has been deposited in the frames or moulds for some time, the latter can be removed and refixed for a further upward portion of concrete. The amount of time necessary for the frames or moulds to remain must depend on the nature of the materials, state of the weather, and thickness of the walls, but even with quick setting cement and a porous aggregate, at least twelve hours should be allowed, and with lime as a matrix, twenty-four hours, if possible, should be the minimum, especially in cold or rainy weather. The fact that frames can be removed without the walls tumbling down, is not evidence that they have unsustained injury, for the effect of withdrawing the supports previous to the concrete having sufficiently hardened is often to split the walls into two longitudinal sections, and which fracture cannot be always detected ; and although with a large margin of strength no disastrous consequences may arise, yet the fact remains that the walls, in places, are in two thicknesses, devoid of bond, and thereby deprived of a large percentage of their strength.

When the panels, whether only rough boards, or

purpose-made wood or iron frames, are liberated, they should at once be scraped, to remove the portion of lime or cement—as the case may be—adhering thereto, otherwise it creates an affinity with the successive deposit of concrete. Some builders apply a soap lather or oil to the faces of the panels each time they are removed, to prevent the adhesion, but in wall-building it is scarcely necessary, if they are cleaned each time of removal. A scraper with a handle about three feet in length is suitable for the purpose, and what is known as a ship's deck scraper also answers effectively.

Packing—that is, stones of considerable size, old bricks, paving-stones, fragments of rocks, large flints, and similar substances—is sometimes embedded in the concrete, with the view, as before stated, of economizing cost, and disposing of old or otherwise useless materials. When walls are not excessively packed, this is a desirable way of accomplishing these objects, but it is common, especially in contract work, to employ too much in proportion to the amount of concrete. Walls under 9 inches in thickness should not be packed at all, and for all others there should be 4 inches of concrete between every course of packing, or if the latter is irregular in form, then no two pieces should be at any one point less than 4 inches apart ; nor should packing *débris* be nearer than 2 inches to the face of any wall. It must also be well embedded in the concrete while the latter is in a soft state, not merely laid on it, or the result may be that the rain will find its way through the irregular cavities that may exist between the bed of the packing and the concrete.

Smoke flues are sometimes formed by inserting taper, or collapsing wood or metal cores in the chimneys during construction, and withdrawing them as the buildings progress. If the aggregate is suitable this method will answer very well, but if it be composed of flint or flint gravel, or anything of a similar nature that will calcine, it is desirable to use terra-cotta or common red ware flue pipes, either socketed or unsocketed. The extra cost of so doing is so small and the advantages so apparent, that it would be well to make a practice of always employing them, no matter what the aggregate may consist of, as perfectly smooth flues are thereby obtained without pargetting, and danger from fire through carelessness in inserting ends of joists, rafters, or other timbers into, or too near the flues, is provided against. For ordinary house flues redware pipes answer sufficiently well, but where exposed to considerable heat the terra-cotta or fire clay tubes, made expressly for such purposes, withstand the action of fire better. Pipes nine inches in diameter are of sufficient size for fire-places in the lower rooms of houses under three floors in height, but above that height the tubes of an elliptical section measuring 14 × 9 are preferable. Where no pipes are used, the flues should be carefully pargetted in the usual manner with hair mortar and cowdung, or with Portland cement and sand.

Fig. 20 is a method of covering over the opening above the chimney bar with ordinary sheet iron, which has a hole cut out in the centre of the size of the flue pipe, and on which the flue is commenced. The advantages of this arrangement are that the sheet iron supports the concrete when in a soft state, without the

necessity of temporary wood supports, and the concrete is afterwards protected from the action of the
fire beneath.

When the wall gauges, tie bolts, wood cores,
or whatever is used to connect together temporarily the wood panels or boards are withdrawn, the
result is a number of holes or cavities in the walls

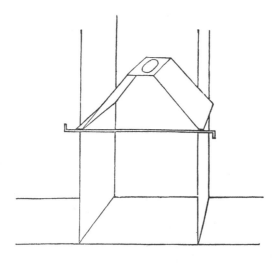

Fig. 20.

requiring to be filled up with cement. These should
never be left unstopped till the walls are about to be
plastered, or probably some may be overlooked, and
thus another way of conveying rain or dampness into
the interior be produced; but as soon as convenient
after the walls are built the cavities should be plugged

with cement and sand (two of sand and one of cement is a fair proportion), not merely stopped superficially with a trowel, but pressed in with a taper piece of wood till the holes are solidly filled.

No lintols are required in ordinary concrete buildings; doorways and other openings must have boards arranged so as to keep the concrete in position till it has hardened, and the heads or soffits must be temporarily supported by a cross piece of wood and props beneath. This method of forming openings for doorways, &c. is so simple and self-apparent as to need no further description. One of the advantages held out by the patentees of concrete building appliances, of which scaffold brackets, or brackets to support scaffolds for the workmen while filling the frames, form a part, is the great saving arising from no independent scaffold being needed. Except in special cases this bracket system of scaffolding should be avoided, and for the following reasons : 1st. The great strain thrown upon the walls at the time they have not attained 20 per cent. of their ultimate strength is an unfair way of dealing with them, and which no other form of building is subjected to under similar conditions. 2nd. The scaffolds are, as a rule, for the sake of convenience of construction, attached to the external sides of the walls of houses and other buildings, and the latter are thereby deprived of the support of division or cross walls, but instead, the scaffold acts as a lever to force them apart. 3rd. As the scaffold brackets are attached to the walls themselves or to the standards, it is quite impossible to finish the surface of the said walls, whatever kind of finish that may be, unless they—the brackets—are removed, and an independent scaffold becomes, after all, therefore a necessity. 4th. In most buildings the work can be executed, as far as construction is concerned, from the inside, with the assistance of tressels resting on the ground or on the floor joists of the upper stories; but as there is no rule without exceptions, so a bracket scaffold may sometimes be used to advantage, and without injury to the walls, and as brackets are usually supplied with all patented appliances, they may be considered as occasional useful auxiliaries. Wood bricks may be, as in any other building, a necessity for fixing the joinery thereto, and for other purposes; these should be of small dimensions, or they may swell with the water contained in

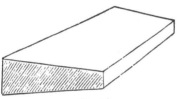

Fig. 21.

the concrete (causing some slight cracks in the latter), and afterwards contract and become loose. Three inches in width by half an inch thick on the lesser or exposed edge, and an inch on the thick edge is ample, as Fig. 21.

As a rule, the less wood built into concrete walls, as into all other walls, the better; and the time-honoured custom of introducing certain timbers of one recognised size, as, for example, wall plates four inches by three inches—the only advantage in this instance being that they occupy the same space as a course of ordinary bricks—may be advantageously abandoned in concrete buildings and smaller scantlings substituted, and, in fact, for floors, wall plates are scarcely required at all if the joists are built in solidly and securely. The practice of leaving cavities in the walls for floor joists, and other timbers is a custom more honoured in the breach than the observance; the idea that the ends would swell sufficiently to cause a rupture in the concrete if built in as the work progresses is not established in practice except where the timbers are of large dimensions, and in such cases they may have their ends immersed in water previous to being fixed, or covered when in place with some weak lime mortar which will allow of sufficient compression to prevent any displacement of the concrete.

Ventilating flues may be readily made in concrete buildings by means of wood or metal cores, or the insertion of common redware flue pipes, or ordinary rain water pipes, in the chimney stacks, or in the walls themselves if of sufficient thickness; and "dead work," or those portions of buildings that fulfil no important duties, but serve only to form the contour of certain parts, as in chimney breasts and stacks, blockings for piers and recesses, &c., instead of being built solid may have "dummy" flues left therein to economize materials and reduce the load.

This practice has at times been suggested as a means of forming water channels, or of embedding therein the soil pipes from closets, waste pipes from sinks, &c.; but the impossibility of ever obtaining access to them in case of a stoppage or for other purposes, render such practices, without doubt, most undesirable. The system of burying pipes of all descriptions out of sight, within walls and under floors, for the sake of appearance, is one of the great evils of modern house construction. It is most essential to build in all door and window frames, for the soundest work is obtained by so doing, and at a saving of time and money. If of an elaborate character, the frames can be cased with thin boards for protection. Great care should be taken that the frames are well " stayed " or " braced " so as not to be easily forced out of their correct shape by the deposition of the concrete round them. Where only a certain portion of the walls can be erected at one time, they cannot be treated in the same manner as if built with bricks, viz. by leaving "toothings" or by "raking back," but indents can be left in them by inserting wedge-shaped cores of wood, so as not to form the said indents of equal depth. Hoop iron can be built in and left of sufficient length to tie into the walls that are to be hereafter built. If all walls can be erected at one time it is far the best plan, but where otherwise it cannot be considered a serious matter.

CHAPTER XII.

THERE are many incidental purposes for which Portland cement concrete can be economically applied besides the construction of floors, walls, and roofs. Wall copings, pier caps, sinks, water troughs, road curbs, gutter channels, cattle mangers, and other things of a similar nature can be formed without any difficulty, if a mould of wood or iron be provided, the reverse shape of the article required. The moulds should be so made as to permit of the concrete being easily liberated when sufficiently hard, and it is much the best plan to construct them so that they can be taken apart,—by means of pins, wedges, bolts and nuts, or some similar way. The manner of making the moulds must depend entirely on the purpose for which the castings are required; but so that they represent an exact counterpart of what is wanted, and will permit of the casting being readily withdrawn, nothing more is needed, and any ordinary carpenter is able to construct them.

It is desirable to roughly plaster the face of the moulds with neat Portland cement about half an inch thick, so that when the concrete is deposited thereon,

and the castings afterwards withdrawn, smooth surfaces are the result; and to prevent the cement
adhering to the mould, the latter should be slightly
oiled each time of use. For any of the purposes
named, the concrete should be composed of about 4 or
5 parts of aggregate to 1 part of cement; a small
quantity of water only must be employed in the mixing,
and the materials be subjected to as much impingement as circumstances will permit. If the moulds
are made with smooth and even surfaces, the faces of
the castings will, when withdrawn, be found to equal
worked stone in appearance, and if the materials employed are of the best quality, and the method of
construction adopted be as here described, the result
will be great durability with immense hardness and
resistance to rough usage, at a cost ranging from one-
third to one-tenth of worked natural stone. Road
gutters produced in the way described have proved to
be capable of resisting the effects of heavily loaded
waggons passing over them.

For horticultural purposes concrete is especially
suitable; the walls of forcing houses may be made one-
half the thickness usual with bricks, for the latter are
as a rule quite incapable of withstanding climatic
changes if less than 9 inches in thickness, whatever
the situation may be, and a brick wall also absorbs
artificial heat to a far greater extent than one of
concrete. For garden boundary walls concrete is a
favourable material, as it remains dry and warm
when a brick wall is damp and cold, and affords
facilities for the fixing thereto of wire or wood strips,
for supporting fruit and flower trees, in a far better
manner than bricks permit. Instead of plugging the

holes in the walls, where ties or bolts have been used as described in page 139, these cavities should.have strips of oak a few inches in length inserted therein, fair with the face or surface of the walls ; these strips should not be driven in too tight, or from the effects of rain and dampness they will swell, and tend to force the concrete apart ; and if the plugs are of dry wood, they should be laid in water for a short time to cause them to expand previous to use.

For garden walls it may be fairly assumed that three-fifths of the thickness necessary if built with bricks and mortar, will be sufficient for concrete of equal strength. Where trees are in proximity to proposed garden walls, and it becomes necessary to provide against growing roots, this may be accomplished by building piers of the thickness of the intended walls and directly under them,—each about 2 feet in length and 8 or 10 feet apart. Instead of digging a deep foundation for the walls themselves, merely a few inches below the surface is then sufficient, as the piers will sustain them ; and if the former are built with Portland cement concrete, there is no probability that they will give way between the supports. If a pier is afterwards in the way of a growing root, the latter will pass round it, and the wall itself will sustain no injury.

For garden walks, or foot-paths, concrete is one of the cleanest, cheapest, and most durable of materials. Unlike tar or the commoner kind of asphalte pavements, it does not "give" in hot weather ; avoids the necessity and cost of weeding and raking, so objectionable a feature with gravel walks ; being smooth it affords easy walking, is less laborious for the wheeling of garden barrows thereon, and every shower of rain tends to keep it clean. To construct concrete paths, edging pieces of wood one on each side, and fixed to plugs driven in the ground, are necessary ; the depth of these pieces of wood must be regulated by the thickness of the concrete, as their top edges must be the guide for the finished surface of the walks, but, as a rule, 2 or 2½ inches thick, if the soil beneath be solid, is ample for most places. It is immaterial what the soil be composed of, on which the concrete is to rest, provided it is hard and fairly dry : if, however, it is loose it will work up, and become mixed with the concrete while the latter is being laid, and if very wet, the water or mud will rise and intermix with it; in either case inevitably destroying the character of the concrete, and rendering it rotten and useless in a very short time. If the ground be dry and loose, it should be rammed, a little water being added to consolidate it, and the garden roller afterwards passed over. When the latter can be used without the soil adhering to it, the path is in a fit state to receive the concrete. If the ground be wet or muddy, some broken stones should be spread over, previous to the deposition of the concrete. Ordinary aggregate of any description may be employed in the proportion of 5 to 1 of cement ; cinders from boiler furnaces may often be obtained, and are suitable, especially if mixed with gravel or other aggregate somewhat harder ; a small quantity of water only for mixing must be used, and the materials well beaten in place with a turf-beater. The wood edging-pieces and plugs may be removed the day following the execution of the work, and refixed for a further portion ; but the surface of the concrete should remain untouched for two or three days, when it should be treated in the way described for floors on page 110 with a mortar composed of half sharp sand and half cement, trowelled off to the surface of the concrete. Care must be taken not to trample or wheel barrows on the finished paths until a week after completion. This system of making concrete walks is also applicable for the paths of villages, or even of towns, if the strength and thickness of the concrete is increased sufficiently to withstand the constant traffic : the only difficulty in the latter case is, in allowing the time necessary for hardening before being publicly used, for ultimate hardness is not attained for months after completion.

In executing this class of work in summer-time, or in hot weather, the concrete should be made to set steadily and slowly by occasionally wetting it, and the trowelled surfaces of cement and sand should, as soon as possible after they are finished, be covered with one to two inches in thickness of sawdust saturated with water. Care must be taken—if appearance is an object—not to use oak sawdust, which would stain the cement a dirty red or brown colour.

The same principle of forming paths is also the most economical method of paving courtyards, making floors for cellars, basements, warehouses, cowhouses, piggeries, corn stores, root houses, and all similar places where wood is objectionable. The average cost is about 1s. 6d. or 1s. 9d. per superficial yard, if $2\frac{1}{2}$ inches thick. Important advantages also in its favour for floor construction for the purposes named are, that it is impregnable against rats, mice, and vermin of any description, and is to a great extent non-absorbent, consequently fairly dry and warm.

Where a space exists between the wood floors of buildings and the natural surface of the ground, the latter should be without exception covered over its entire area with concrete in the way just described, but 2 inches in thickness would be sufficient and ground lime concrete would answer the purpose. This would largely conduce to the healthfulness of buildings, and assist in preventing decay of the floor timbers.

Underground water tanks for storing rain-water from roofs of houses for domestic purposes, farm buildings and stables for the supply of soft water to cattle, garden buildings for the many purposes for which it is there required, manufactories for trade uses in preference to hard water, and in many other instances where a storage of soft water is an important consideration, can be made of concrete at about half the cost of that of brick and cement. The homogeneity of concrete, and its easy application, render it especially suitable for works of this description. Circular or oval-shaped tanks are necessarily the strongest and most reliable ; these may be made by using an ordinary close boarded wood drum 4 or 5 feet in height, between the perimeter of which and the excavated ground the concrete is deposited ; when the concrete has sufficiently hardened, the drum is liberated and re-fixed immediately above the part already executed, and so on until the full height is reached. The drum must be made in two or three sections, to facilitate removal ; if this is not done the concrete will press against it and prevent its release except by being taken to pieces. The

bottoms of concrete tanks should be executed, to avoid
injury, after the sides are completed. The tops may
be covered with concrete arch-shaped, on wood ribs,
or better still as a slab, on flat boards, the latter
having a bearing on the walls of about 2 inches.
These boards can be left in, and the saving of labour
and concrete materials by this method compensates
for the loss sustained by the boarding not being re-
movable. An economical size and shape for water
tanks is 12 feet in length, by 6 feet in width in clear,
and elliptical on plan, and for a tank of this description
not more than 15 feet deep the walls and bottom
should be 12 inches thick, the top being a flat slab of
concrete on boards as described 6 inches in thickness,
with man hole if required left in the centre in the
usual manner. The inner surfaces of the tank should
be either rendered in Portland cement and sand half
of each and three-quarters of an inch in thickness, but
laid on and finished in one coat, or well trowelled to
the surface with neat cement in the way described as
suitable for floors on page 110. The latter method is of
course the most economical, and in a considerable number of cases has never failed. Where average facilities
exist for obtaining the necessary materials, a tank of
the size named should be constructed complete, excavation included, for about £22, and would contain 5,400
gallons of water. This estimate is for walls and
bottom of Portland cement concrete, in the proportion
of 7 parts of aggregate to 1 part of cement, and 12
inches in thickness, the top a flat slab 6 inches thick,
and the inside trowelled with neat cement as described.
The cost per thousand gallons would thus be about
£4 or thereabouts, but a saving of £3 on the

total amount may be effected by using as a matrix,—

3 parts by measure   ground lias lime.
1 ,,   ,,   ,,   Port and cement.
24 ,,   ,,   ,,   aggregate.

This formula has been adopted in many cases, and for some considerable time, with the best results. The wood core or drum for forming a tank of the size and description named is necessarily heavy and clumsy, but a patented appliance made of iron in sections to allow of easy removal, and which is also adapted for tanks of different sizes, can be obtained for about £6. This may be fixed by two ordinary labourers in about two hours (see page 92).

Stairs may be constructed in the ordinary manner usual with stone,—by casting each step separate in a mould, and pinning them into the walls at the ends; or forming one block by means of rough boards in the line of the soffit, and others also temporarily fixed to enclose risers and treads. Indents should be left for the wall ends of steps in the walls during construction, and a light iron girder is best for supporting the open ends when the stairs are unsupported on one side. The treads and risers may be finished with Portland cement, or the risers lined with tiles, and the treads have an oak nosing attached, with blocks of wood or tiles at the back, as shown in the details of construction for a concrete villa at the end of the volume.

CHAPTER XIII.

The cost of concrete, more especially in the construction of walls, has been, from its first introduction, one of the points on which opinions have varied to no inconsiderable extent. On the one hand we have a statement put forth in a pamphlet that it is possible to execute concrete walling for 4s. per cubic yard, and on the other hand, we have had repeated public assertions that the same amount cannot be built under 15s. to 18s., and also the undeniable fact that alternate tenders for buildings in brick and concrete have resulted in favour of the former material—for cost. But it is only fair to say that instances of the latter kind have been exceptional ones, and for which there may have been exceptional causes. Like brickwork, and in fact like all other items of building, the cost of concrete depends on various circumstances. There may be good local facilities for obtaining the aggregate—sometimes possibly, gravel dug out of the foundations or basements of the intended buildings, and ready for immediate use, at other times rough stones or boulders may have to be carted some considerable distance, and crushed by steam power.

Cement or lime may be procured within a short distance of its destination, or have to be carried by railway one or two hundred miles, and be then a day's journey possibly from the railway-station to the building. Water is perhaps available close to the surface of the ground, and, on the other hand, may have to be carted in barrels.

In estimating for concrete work, therefore, there must be considered—

1st. The cost of each constituent delivered on the spot, ready for use, and their relative proportions.

2nd. The supply of water.

3rd. The character of the building, whether plain or intricate on plan, if blockings for cornices or projections for windows or other irregularities occur in the elevations, and the thickness of the walls (thin walls costing more in proportion to their bulk than thick).

Relative to the amount of water required for mixing concrete, and where water is not abundant it is therefore of some consideration, Mr. Homersham, C.E., in the *Journal of the Society of Arts*, says : " The minimum cost of burning ballast when clay fit for the purpose is found on the site of the works, may be put down at 3s. per cubic yard, including the cost of labour in sifting and washing, but not of providing the necessary quantity of water. The quantity of water required for washing and soaking burnt ballast is about 20 gallons per cubic yard, or a ton weight per rod. One half that quantity will suffice for washing good gravel, containing the proper proportion of clean, sharp, silicious sand ; such only should be used in making con-

crete.   The quantity of water requisite for gauging the
compo is about three gallons per bushel of Portland
cement, and one gallon per bushel of damp sand."

As concrete walls, unlike brick ones, can be built of
any required thickness, varying in single inches, it is
usual to measure and estimate for their construction
by the cubic yard.   The practice is simple in measurement, easily calculated, and universally understood,
and the reduced standard rod, square, local perch,
rood, rod, rope, and endless other terms and forms
of wall admeasurement, it is to be hoped will eventually be ousted in favour of the much simpler cubic
yard.

In estimating the cost, it must be remembered that
a diminution in bulk of the aggregate occurs during
its formation into concrete, and that the amount of
diminution depends entirely upon circumstances ; for
instance, Thames ballast in a wet state with its particles
pressed together by the weight of water it contained,
would clearly lose less in bulk than dry pit gravel,
light cinders, burnt clay, or a similar description of
aggregate.   With either of the latter materials the
concrete would probably measure 20 per cent. less than
the aggregate (the cement or lime in no way increases
the bulk), whereas wet Thames ballast would not lose
more than 5 per cent. of its volume.   It is a safe
method in estimating for ordinary buildings, taking
thin and thick walls together, to allow the space
occupied by packing to compensate for this diminution
of bulk, although in many cases, with only a fair
amount of wall packing, the latter considerably more
than makes up in measurement for the shrinkage of
the concrete.

The cost of labour should vary but little in country districts,—in towns it would depend upon the ordinary rate of wages; but, as a rule, for walls of average height and thickness in ordinary house construction, 2s. would be the minimum and 3s. the maximum value of mixing and depositing the concrete in place, including pumping water and the erection and striking of bracket, tressel, or other form of scaffolds.

The cost of fixing the temporary appliances or moulds for defining the walls, exclusive of the value of fixing the joinery, or other timber or ironwork that may be necessary to insert in the walls during construction, would average from 6d. to 1s. per cubic yard of finished walling. The more simple the plan of the erection, the greater facility, and consequent less cost, for fixing.

A fair sum to allow for hire, or use and depreciation of the appliances for monolithic construction, is 6d. per cubic yard if the works are of any magnitude, and more for small buildings. With these data, the probable cost of concrete building may be arrived at with tolerable accuracy; taking imaginary prices for the materials and labour, the proportions being 7 parts of an aggregate to 1 part of Portland cement, and, for sake of comparison, the probable prices of bricks, lime, and sand, we should get the following results :—

CONCRETE.	s.	d.
1 cube yard of clean gravel, fit for use, delivered at	4	0
2¾ bushels of Portland cement delivered at 2s. 6d...	6	7
Labour of mixing and depositing	2	6
Do of fixing appliances	0	9
Use and depreciation of do.	0	6
Per cube yard......	14	4

Here the packing is assumed to compensate for the shrinkage in the concrete itself.

	s.	d.
BRICKWORK.		
400 bricks at 40s. per M delivered	16	0
⅛ of a yard of lime at 12s.	1	6
¼ of a yard of sand at 2s.	0	6
Labour	4	0
	£1 2	0

In deciding the thickness of external walls to be built of concrete, good results have been obtained by adopting 9 inches for cottages, even in exposed situations, and 12 and 14 inch for buildings where 14 and 18 inches respectively would be the thickness if bricks were employed. With these comparative dimensions, concrete is much superior to ordinary brickwork, both for strength and dryness, so that the actual cost of the former should fairly be credited with the saving in quantity, and not compared for cost with a material of equal thickness—but of equal strength. Concrete walls, however, require some degree of external finishing, which common brickwork does not; but the latter necessitates "gauged" or common arches, or stone lintols to external window and door openings, and wood lintols or discharging arches for internal openings, which together are a fair set-off against a plain way of finishing concrete walling; but if something more than plain finishing is a requirement for a concrete building, it may be assumed that something more than plain brick-work would be wished if it were a brick building.

Mr. Homersham, C.E., in the *Journal of the Society of Arts* says: "The cost for labour of all kinds (concrete labour) should not exceed 2s. 9d., but say 3s., per cubic yard, or £1 14s. per reduced rod; and the case will prove to be very exceptional where the cost of Portland cement concrete *in situ* in a building proves to be on the average less than 10s. or more than 16s. per cubic yard."

The Hastings Cottage Improvement Society in their annual report for 1873 state, with regard to their new buildings: "There are eighty-three apartments, divided into thirty-one tenements. The cost of building was £2,961; the site is valued at £350; and other items of expenditure (including an allowance for interest on unproductive capital during erection) make the total outlay about £3,580. The estimated saving effected by using concrete instead of brick, is £406, or about 15 per cent."

Mr. Tall, in the *Builder* for June 10th, 1871, says he "has erected concrete walls costing from 6s. to 12s. per cubic yard in different parts of the country;" in this case, the most unfavourable district must have been anything but deficient in the means of economic concrete building.

Mr. Wonnacott, in a paper read at the Architectural Conference in 1871, stated as his opinion that the cost of concrete building might be estimated at from 12s. to 18s. per cubic yard, and the following paragraph on the same subject is extracted from the *Building News* for December 13th, 1872:—"This statement is furnished us by Mr. Stewart, Lord Kinnaird's Surveyor and Clerk of Works, of the cost of concrete building of the parsonage at the Knap near the Mill Hill Lodge, Rassie Priory, Perthshire, when 10 feet high from the bottom of foundations:—

"CONCRETE.—Measurement of main division and partition wall 18, 12, 9, and 6 inches thick, reduced to a standard of 12 inches thick, voids deducted—130 superficial yards. Expenditure :—

Cement used, 4½ tons, at 50s.; in proportion of 8 of small material to 1 of cement......	£11	5	0
Labour, 37 days, at 2s. 6d.	4	12	6
Extra for foreman... ·		2	6
Joiner for framing, &c., four days, at 4s.		16	0
	£16	16	0

" £16 16s. divided by 130 gives 2s. 7d. as the cost of one yard superficial for cement and labour ; or, for comparison, £4 13s. per rood of 36 square yards. Materials used in 130 yards superficial of concrete wall 1 foot thick are 32 loads small material, 14 loads large material for packing.

"MASONRY.—In measurement of masonry, no voids under 8 feet in width are deducted to allow for plumbing, semilions, sides, &c., but all deducted in this case. Measurement as before, 130 superficial yards in walls 22 in. thick :—

130 yards nearly equal, 3⅗ roods of 36 square yards, 3¾ roods, labour only, at £6 5s...........	£22	18	10
Cost of concrete...........................	16	16	0
Balance in favour of concrete in labour only	£6	2	10

"The above does not include cost of materials nor cartages. In order, therefore, to show the difference or quantity of materials, I will take the cartage and put in a price similar to both:—

CONCRETE.

Labour only	£16	16	0
32 loads small material at 1s.	1	12	0
14 loads larger at 1s.		14	0
	£19	2	0
4½ tons cement from station..............		18	0
Concrete—Total labour and cartages ...	£20	0	0

Masonry, 22in. thick—

3¾ roods, 87 loads, at 1s.	£4	7	0
3½ tons lime from station, at 4s.		14	0
28 loads of sand, at 1s. 1		8	0
Cost of labour only...........................	22	18	0
Masonry—Total cost £29		7	0

Cartages, of course, vary according to distances.

"The cost of concrete for labour only and cement, 2s. 7d. per yard. The cost of masonry, labour and lime only, 3s. 6d. per yard."

In an experimental case in Wiltshire, a pair of labourers' cottages were built with rough stone, some new bricks being used where necessary. The external walls were 16 inches thick, and the stone itself was on the spot free of cost, and subject to no charges for haulage. A pair of concrete cottages were built adjoining these and at the same time, the walls being 9 inches thick, and composed of 1 part of Portland cement to 7 parts of river gravel. The latter had to be carted 3½ miles, the total charges when delivered being 4s. 6d. per cubic yard; the cement cost 60s. per ton delivered on site, and the apparatus for constructing walls was charged at 6d. per cubic yard of work executed, for use and depreciation, besides a considerable sum for conveyance. The works were all executed by measurement, and, with the exception of the walls and chimneys, all other portions of the buildings were similar, and executed by the same workmen at an uniform scale of payment; the internal dimensions and arrangements were also equal. With the heavy handicapping the concrete cottages were subject to, their ultimate cost was less by a few pounds than the stone cottages, with the major part of their materials for walling on the spot free of cost.

The cost of concrete, when compared with other building materials, must, however, be so conditional that no definite comparison can be made, and it is not even suggested that it can always be employed with advantage, for there are certain descriptions of buildings for which it is more specially adapted than others, and in no instance is this so conspicuous as in the class of erections necessary on landed estates. For farm buildings, labourers' cottages, &c. that are now required to be of a more superior character than formerly, but which give a money return quite inadequate to their cost, concrete is especially suitable; for unsheltered as the majority are by adjoining buildings, exposed to strong winds and drifting rains from all points of the compass, nothing short of hollow walls are damp-proof, except concrete. No other suitable material can be used that will find employment for the unskilled and surplus labour of

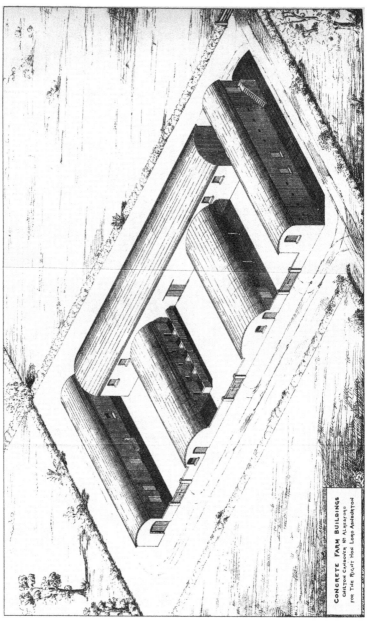

CONCRETE FARM BUILDINGS
CHILTON CANDOVER NR ALRESFORD
FOR THE RIGHT HON. LORD ASHBURTON

The material originally positioned here is too large for reproduction in this reissue. A PDF can be downloaded from the web address given on page iv of this book, by clicking on 'Resources Available'.

rural districts. It is adapted for water tanks, lining of ponds, troughs, and many other purposes where cattle are kept,—is almost invulnerable to vermin,— is durable,—will resist any reasonable amount of rough usage,—and last, but not least, for the class of work named, it can, in nine cases out of ten, be applied for half the cost of brick walls of equal strength.

On many estates there are, unfortunately for the owners, a large proportion of old buildings past repair and unfit for habitation or occupation, and the materials they are constructed with are in many instances apparently useless for any purpose whatever, but by the aid of a steam crusher and Portland cement, old brickbats, flints, stones and tiles are indiscriminately converted into a material every way qualified for the re-construction of farmsteads. And the plans of such buildings are, or should be, necessarily of the plainest character, free from irregularity or intricate design, giving ample space, but avoiding those peculiarities of construction that engender a larger outlay than the profits of farming will permit. Cheapness, strength, and durability should be the objects sought for, and monolithic concrete with our present knowledge of its use and the means of applying it, exactly meets these requirements. The illustration of Chilton Candover farm buildings is a representative building of the kind described. Erected on a site adjoining old buildings, the *débris* from the latter, brickbats, stones, &c. are quickly passed to a Blake's crusher and transformed into an aggregate of very suitable character. The upper floors are principally concrete, resting on iron girders, and the tanks, water-troughs, and some portion of the pavings

M

are also of the same materials. The composition of the concrete for walls is—1 part of a matrix composed of blue lias lime and Portland cement, to 7 parts of aggregate; and the relative proportions, by measurement are—

21 bushels of aggregate.
2 ,, ,, blue lias lime.
1 ,, ,, Portland cement:

and the actual cost does not exceed 13s. 7d. per cubic yard, as follows—

	£	s.	d.
Pulling down old buildings, removal to crusher, crushing, including value of depreciation of engine and plant	0	3	0
2 bushels blue lias lime, delivered on works, at 11d.	0	1	10
1 bushel Portland cement, delivered on works	0	3	0
Labour of mixing and depositing concrete	0	2	6
Labour of fixing appliances, and removing same.	0	0	9
Use of appliances and depreciation ...	0	0	6
Rough trowelling both sides of walls ...	0	2	0
Cost per cubic yard—complete	£0	13	7

The external walls are 12 inches thick, and comparing these with brick walls 14 inches thick,—which would, however, be inferior in every respect—at the local prime cost of £14 15s. per reduced rod, or £1 6s. 3d. per cubic yard, the result would be a difference in cost of about 125 per cent, or what can be had in superficial area for 13s. 7d. in concrete, would be £1 10s. 7d. in brickwork. This, it must be admitted, is an exceptional case, the aggregate costing but little, and bricks of good quality being dear.

CHAPTER XIV.

THE EXTERNAL TREATMENT OF CONCRETE BUILDINGS.

THE proper and legitimate finishing of concrete buildings externally—except in cases where appearance is
of no object and the practice of rough trowelling previously described (page 110) is practised—is, and
always has been, since its introduction for house
building purposes, a "bone of contention." Even
this rough trowelling, however, with a narrow band
of cement, projecting about half an inch from the
surface of the walls to all quoins of buildings and to
window and door openings, is far from unsightly, and
when the colour is toned down by time, has anything
but an objectionable appearance, looking, what it
really is, a monolithic construction.

With concrete block building the treatment of finish
would be the same as to a stone or brick building,
and therefore needs no further remarks.

As before stated, the difficulty with appliances for
forming walls on the continuous system, is that they
afford no means of introducing projecting surfaces,
without a very considerable extra amount of trouble,
labour, and delay, and therefore whatever relief is
intended to be given to a concrete front where Port-

M 2

land cement is objected to, seems to point to some system of sunk panel, or flat surface ornament.

Sgraffito in panels might be advantageously used, and so could tiles in bands or panels, and the small depth of sinking in walls of ordinary thickness necessary for these, would have little or no prejudicial effect on their strength. Small tiles, or even mosaic work, might be introduced by arranging the materials face downwards in shallow boxes of the exact size of the panels, and filling up the boxes with fine concrete: the tiles and this backing of concrete need not be together more than an inch or an inch and a quarter in thickness; recesses in the walls during construction would have to be left to afterwards receive the panels.

For window heads, string courses, friezes, &c. some such system might be inexpensively and satisfactorily adopted. It has been attempted to introduce split flints as a finish to the flat surfaces of walls, by embedding them in the concrete during construction, but it has practically failed, not because it was an unfitting treatment, but through the difficulty of keeping the flints in position, in a soft spongy mass like concrete until it has hardened. Ashlar stone dressings, and faced flint surfaces or panels with a concrete backing, would for a country residence be a species of wall construction not easily surpassed, were it attainable at a moderate cost and a reasonable amount of labour and patience.

The introduction of bricks for quoins and dressings has also as a rule proved unsuccessful, although desirable, were it fairly practicable.

The practice of colouring or staining cements, fit for external work, in a soft state, and previous to their application, does not appear to have arrived at a decided stage of success. Probably the nature of Portland cement renders it difficult to deal with in this respect without injuring its quality, for it does not maintain uniformity of colour, eventually becoming " spotty."

Round or oval pebbles, inserted in a rough or rendering coat of cement on the walls after the appliances have been removed, is for country building an appropriate finish, especially if arranged in panels, with a margin fillet or band of cement. Rough dash, or rough cast, is also an inexpensive way of dealing with the walls of cottages, lodges, &c.: this may be done by mixing fine clean gravel and course sand with slaked lime, or lime " putty," or cement by preference, into a semi-fluid condition, and, as its name implies, casting or dashing it on the surface of the walls; but the quoins should have a neat fillet of cement, or of some other material, for forming a sharp angle or arris, or they appear irregular and ragged. Rough cast may also be executed by applying a rendering coat of cement and coarse sand, in the proportion of 1 of the former to 3 of the latter by measure, and throwing shingles into it when in a soft state. In this case the shingles should be of an uniform size and washed very clean, and not too heavy, or it will be difficult to make them adhere. This is very suitable for buildings where shingle of a good colour, or of various colours intermingled, can be · procured, and offers, in conjunction with tile strings or bands, and cement quoins, an opportunity for appropriate methods of finish.

Stamping patterns on a rendering coat of lime and hair, or cement, or even marking with a trowel, would be an economical way of treating an inferior class of buildings.

Half-timbered erections of the usual descriptions, filling in between the timbers with concrete, and rough cast, plastered, or stamped panels, has been suggested; but these scarcely come within the category of concrete buildings. The objections however are—(1st) the difficulty and cost of obtaining thoroughly-seasoned and suitable timber, and therefore the possibility of the wood shrinking away from the concrete, and admitting rain and moisture (2nd) the application of the concrete in small pieces rendering it costly, and troublesome to execute, and (3rd) making the duration of the building dependent on a perishable material, and the most lasting of the two materials subservient to the inferior and weakest.

This may be remedied, however, by building the wall with concrete in the usual manner, and inserting small blocks of wood in the soft concrete, for fixing thereto a facing of timber about ¾ inch thick, so that in place of solid timbers they would be "veneers." The frontispiece, which is a reproduction from the original drawings of a design for a concrete villa, and which was awarded the first prize by the proprietors of the *Building News*, shows the application of the principle suggested, and the details at end of volume explain the method of treatment. The advantages of this method compared with solid timber erections are—strength and durability of a monolithic concrete building; easy application and economy of cost; impossibility of damp penetrating, consequent dryness of walls, and durability of the timbers; facility for renewing the woodwork when decayed, and which would be scarcely more difficult than replacing the weather boarding of a barn or the boarding of a floor, and equality in appearance to solid timbers. The objections—not disadvantages—raised against the system would perhaps be—want of thoroughness, and to a certain extent a pretence of being what it is not; but this can scarcely be a heavy charge, for no attempt need be made to give it an appearance different from what it really is, viz. wood, with rough cast (on concrete) panels; and the small quantity of timber required would permit of seasoned oak being used and left unpainted. Where solid timbers might be employed, the probability is that in the majority of cases Baltic or Russian fir timber would be employed, and afterwards painted and grained to imitate oak. In all such cases, the latter way of building must surely be the greater sham of the two. On a small scale, the way suggested of timber facing to a concrete building has been tried, and realised, as far as cost, appearance, and facility for construction was concerned, the success that was fairly anticipated.

The best way of finishing externally concrete buildings is still open to discovery, and probably no one will be found to deny that success on that point has not kept pace with the manner of dealing with it as a structural material.

CHAPTER XV.

In the interest of all concerned with the future use of
concrete for building,—of employers who venture on
seceding from the ordinary methods of construction
and of architects whose reputation may be imperilled
by the improper executions of their designs,—it is not
only necessary to obtain good materials and procure
efficient surveillance, but, where buildings are to be
erected by contract, that a complete specification of
the duties of the concrete workmen should be defined.
Not a small amount of at least unpleasantness, and
possibly, litigation, may ensue from incompleteness
in this respect. These remarks may appear ambi-
guous, in the face of the generally clear and explicit
specifications that accompany the majority of architectural drawings, but it is well known that, probably
through dealing with a material which is not generally recognised, and its capabilities therefore only
partially understood, the definite instructions and
conditions accorded to the concrete workmen have,
unlike those of other occupations, been of the most

meagre description. To this omission may be traced the unsuccessful issue of certain concrete buildings, which, although not positive failures, cannot be credited as favourable specimens of a material from which so much was anticipated. In the construction of one of the largest houses hitherto built with concrete, and under the superintendence of an eminent architect, the stopping of the holes in walls caused by the ties or wall gauges, the parging of smoke flues, and other incidental portions of the work, were matters that occasioned continuous disputes, because the contractors disavowed any obligation for the performance of those very necessary portions of the work, through the absence of specific instructions in their agreement. But it will be seen that concrete construction demands as strict details of what shall and what shall not be done as any other branch of building ; and the following specification which would apply to the erection of buildings similar to the one illustrated at the commencement of this work, will serve as an embodiment of the instructions laid down for successful concrete building.

CONCRETE CONSTRUCTION.

The concrete for external and cellar walls is to be composed of Thames ballast or gravel, clean pit gravel, crushed flints, stone chippings, or any other materials of a similar nature, hereafter called " the aggregate," and Portland cement, in the proportions of 7 of the former to 1 of the latter by measure. *Constituents, and their proportions for walls.*

The internal walls, chimneys, soft-water tank, casing of ground beneath floors, foundations to level *Constituents for internal walls,* chimneys, tanks and sleeper walls. of ground, and sleeper walls, 21 parts of aggregate, as before described, 2 parts of ground blue lias lime, and 1 part of Portland cement, all by measure.

Constituents for vaults, stair and chamber floors. For vaults of cellars, formation of stairs, and landings, and chamber floor construction, 5 parts of aggregate, as before described, and 1 part of Portland cement, both by measure.

Size and shape of aggregate. The aggregate will be preferred of an irregular size and angular shape, must be perfectly free from fine sand, and clayey, loamy, or argillaceous matter of any description ; and while a sufficient proportion of coarse sharp sand shall form a constituent, it must not exceed such proportion as would tend to weaken the character of the concrete.

Size of aggregate for vaults and floors. For internal division or partition walls under 9 inches thick, and the floors, cellar vaults, and stairs, the aggregate must not consist of any portion that would not pass through a one and a quarter inch sieve or screen, and for all other purposes through a two and a half inch.

Test for the quality of cement. The Portland cement to weigh not less than 100 lbs. per bushel, and a block of the same $1\frac{1}{2}$ inches square and mixed neat, shall, after immersion in water for seven days, be capable of sustaining a tensile strain of not less than 450 lbs.

Cement not to be used fresh. If kept in bags or barrels, the cement is not to be used within twelve days after delivery on the works, or six days if spread out during that time on a wood floor in a building impervious to rain.

Treatment of lime previous to use. The blue lias lime to be finely ground and kept in open top wood bins or enclosures in a dry building, for at least six days previous to being used, the lime during that time to be occasionally moved or stirred.

The requisite proportion of water cannot be speci- Proportion of water necessary. fied, but must depend upon the relative absorbent nature of the aggregate, and the amount of moisture it might contain previous to use.

For mixing the ingredients, platforms not less than Platforms for mixing the materials. 10 feet square to be provided, and as many as may be requisite.

Not more than two-thirds of a cubic yard of concrete Quantity to be mixed at one time. to be mixed at one time, and proper measures for holding the specific quantity of each material to be employed.

The ingredients are to be turned over twice dry and Method of mixing the materials. twice wet, by not less than four men, one of whom shall use a two-pronged hook or rake to aid in thoroughly incorporating the mass, and another to add the water gradually by means of a watering-pot with a fine rose fitted thereto. The materials to be kept moving during the time the water is being added, to prevent the cement (being the finest portion) sinking away from the coarser.

No more water than is actually necessary to be Less water to be used for vaults, &c. used; and for the foundations, floors, vaults, and stairs, merely sufficient to allow the cement to adhere to the aggregate.

In the formation of all walls not less than 9 inches Filling-in pieces or packing to walls. thick, large stones, brick-bats, or similar materials may be inserted to economize the amount of concrete, subject to the following conditions :—No two filling-in pieces or packing shall be less than 4 inches apart from each other, and 2 inches from either side of any wall, and if of irregular form, the largest side to be placed downwards; all filling-in pieces to be free from dirt or other injurious matter, and to be rubbed or embedded into the concrete, so that they may adhere properly, and no cavity be occasioned in the walls.

Concrete beneath ground-floors.

The entire area of the building beneath floors to have a layer of concrete deposited thereon 3 inches in thickness (but the ground need not be perfectly level for this purpose), and well beaten with a wood beater about 12 inches by 9 inches, similar in all respects to a turf beater, but the face covered with sheet iron (to prevent the concrete from adhering). This layer of concrete is to prevent damp rising, stop the growth of fungi or vegetable matter, and to form a barrier against rats or other vermin.

Grouting, &c. vaults and floors.

For the floors and vaults, the concrete is to be well beaten and afterwards grouted with liquid cement and sand, one of the former to three of the latter, which is to be free from all impurities.

Construction of stairs and landings.

In the construction of stairs and landings, indents are to be left in the partition walls corresponding with the inclination of the former, to sustain the wall-ends of concrete-treads. The iron girders to carry open sides of stairs and landings to be fixed in place and rough boarding temporarily attached thereto, and of the necessary shape and form, to allow the concrete being deposited therein. Build in the wood blocks necessary for fixing tread-nosings thereto.

Fireplace construction.

The fireplace openings to be constructed as shown on plan, and the gatherings over chimney-breasts with sheet-iron plates made thus (see Fig. 20, page 139), for the protection of the concrete from heat of fire. The flat portion or upper part to have 9-inch circular holes cut therein, on which are to be commenced terra-cotta socket flue-pipes to form smoke-flues, and which are to be carried up the full height of chimneys, with the necessary bends to prevent down drafts, &c. Chimney bars made from 1½ by ¼-inch iron in the usual manner are to be used.

The ventilating flues are to be pipes of the same description as used for smoke-flues, but 12 inches below ceiling-line in all rooms right-angle 4-inch junctions are to be inserted, into which 4-inch ordinary socket-pipes are fixed, and which are to finish fair, or flush with the plastering of walls, and in the socket of which pipes common circular iron hit-and-miss ventilators are to be fixed (prime cost 1s. 6d.), and made to open or close by means of suspending cords fixed to knobs of closing portion of same. The ventilating flue pipes are to have 6-inch junction-pipes leading out of same, and finished fair with outside face of chimney stacks, about 12 inches above the ridge, the sockets having perforated iron gratings fixed therein to allow for the escape of vitiated air. The ventilating flues above the junction to be covered over, and not to remain open to the top, as with smoke flues. *Ventilating flues.*

Fresh air to be admitted to all rooms by means of ordinary air-bricks fixed in external walls immediately beneath floor-level, and wood trunks leading from thence to fireplace openings, the iron sides of grates to be drilled or otherwise perforated, to correspond with the position of air-trunks, the air from which will pass through the perforations into the rooms. *Ingress for fresh air.*

Insert one course of 1-inch No. 16 gauge iron hoop every 2 feet in height throughout the entire length of all walls. *Iron-hoop bond.*

No wood lintols over door-way are needed, but, in the case of door or other openings occurring in lime concrete walls, Portland cement concrete shall be used *No lintols necessary.*

not less than 18 inches deep, and 24 inches wider than the respective openings.

Wood blocks for fixing of internal joinery.   Build in internal walls wood blocks for the purpose of fixing door-jambs or other wood-work thereto, of the best red timber or oak and of this section, (Fig. 21, page 141) and wherever required.

Wood blocks for fixing of external joinery.   As the works progress, insert red deal or oak blocks of the same section as used for internal fixing, in the face of the external walls for the necessary fixing of the ¾-inch wood casings, linings, or other timber-work, as shown.

Making good previous to plastering.   Previous to the commencement of plastering walls, either external or internal, thoroughly stop with cement and sharp sand (one of the former to three of the latter) all holes in walls occasioned by the temporary use of wall gauges, bolts, ties, or other portions of concrete appliances.

Centreing, &c., to be saturated with water when concrete is to be deposited thereon.   The centreing, turning-pieces, and other timber work (except timber uprights) required for supporting the concrete in its soft state, for vaults of cellars, door and window heads, chamber floors, &c., are to be thoroughly saturated with water and allowed to expand previous to the concrete being deposited thereon ; otherwise, the work being purposely rigid, to carry the weight of concrete, will swell with the water contained in same, and gradually heave the concrete during the action of setting and hardening, and eventually cause cracks and defects in the walls.

Boarding of vaults, &c., to carry concrete to have open joints.   The temporary boards of cellar vaults and chamber floors requisite to carry the concrete are also to be laid with ¼-inch spaces between them to allow the concrete to part with the excess of water, other than sufficient for hydration. (The water parted with will not carry with it any portion of the cement.)

Hanging brackets to form workmen's scaffolds are not to be allowed, as the concrete workmen can use tressel-scaffolds with equal advantage and avoid the evil of unnecessarily straining walls before they have sufficient time to attain moderate strength. *No hanging scaffolds to used.*

Independent scaffolding is to be employed for finishing the exterior of the building, and no put-log holes are to be made in the walls. *Independent scaffolds for finishing exterior.*

The concrete chamber floors are to be executed when the walls are the necessary height for that purpose, but the floor-joists are not to be fixed until the building is roofed in. *When chamber floors are to be constructed.*

The rolled iron joists are to be of the section shown by detail drawing, and have a bearing on all walls of 9 inches ; no bedstones or templates are necessary. *Section of iron joists.*

All iron joists are to have one coat Torbay oxide of iron paint previous to being fixed, and one coat afterwards, but before the concrete work is executed. *Iron joists to be painted.*

The soft-water tank to be 8 feet diameter in clear, formed round a wood or iron mould—the side, walls, and bottom to be 12 inches thick, the top formed with a slab or layer of concrete 6 inches thick, having a manhole 2 feet square in the centre, and covered with a concrete slab 2½ feet square and 4 inches thick with ring let and bedded into same. *Underground tank*

THE END.

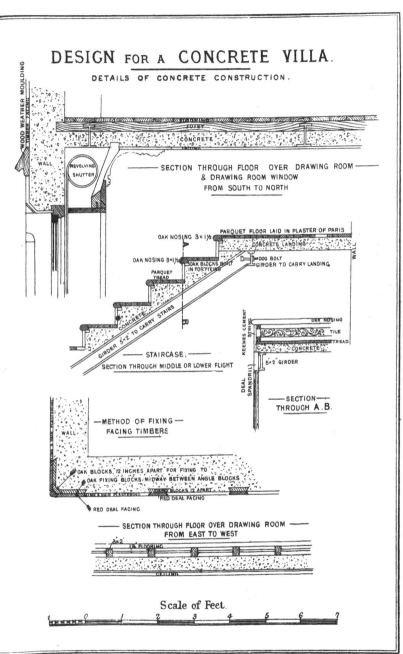

DESIGN FOR A CONCRETE VILLA.

DETAILS OF CONCRETE CONSTRUCTION.

WOOD WEATHER MOULDING

TIMBER FACING

JOIST

CONCRETE

WALL

REVOLVING SHUTTER

— SECTION THROUGH FLOOR OVER DRAWING ROOM —
& DRAWING ROOM WINDOW
FROM SOUTH TO NORTH

PARQUET FLOOR LAID IN PLASTER OF PARIS

OAK NOSING 3 × 1½

CONCRETE LANDING

WALL

OAK NOSING 3 × 1½

OAK BLOCKS BOLT IN FOR FIXING

DOG BOLT

GIRDER TO CARRY LANDING

PARQUET TREAD

CONCRETE

GIRDER 5 × 2 TO CARRY STAIRS

OAK NOSING

KEENES CEMENT SOFFIT

TILE

TREAD

CONCRETE

— STAIRCASE. —
SECTION THROUGH MIDDLE OR LOWER FLIGHT

5 × 2 GIRDER

DEAL SPANDRIL

— SECTION —
THROUGH A.B.

— METHOD OF FIXING —
FACING TIMBERS

WALL

OAK BLOCKS, 12 INCHES APART FOR FIXING TO

OAK FIXING BLOCKS MIDWAY BETWEEN ANGLE BLOCKS

BLOCKS 12 APART

RED DEAL FACING

RED DEAL FACING

— SECTION THROUGH FLOOR OVER DRAWING ROOM —
FROM EAST TO WEST

3 × 2

1½ FLOORING

CEILING

Scale of Feet.

1   0   1   2   3   4   5   6   7

UNWIN BROS. LITH. LONDON E.C.

F. W. REYNOLDS & Co.,
ENGINEERS & MACHINISTS,
73, SOUTHWARK STREET,
AND THE GROVE, LONDON, S.E.

To Her Majesty's War Department;
The Indian Department of Her Majesty's Government;
The Spanish Government Fortification Works
at Cadiz, &c., &c.

*Five Silver and Two Bronze Prize Medals have been awarded
to F. W. R. & Co. for Wood Working and other
Hand and Steam Power Builders'
Labour-saving Machinery.*

SAWING MACHINES, BAND SAW TABLES,
MORTISING MACHINES,
**Tenoning Machines, Boring Machines, Drilling
Machines, Shearing Machines,**
PUNCHING BEARS, MITRE CUTTERS,
GENERAL JOINERS,
HORIZONTAL AND VERTICAL ENGINES AND BOILERS,
STONE SAWING MACHINES,
CONCRETE BUILDING APPARATUS,
MORTAR MILLS, PAINT MILLS, GRINDSTONES,
SLATE HOLERS,
FLOOR AND BENCH CRAMPS,
*Cranes, Crabs. Lifting Jacks, Pulley Blocks, Tire Benders,
Builders' Screens, Forges and Bellows, &c. &c.*

*In purchasing through Agents see that F. W. REYNOLDS
& Co.'s Name and Address is cast on each Machine.*

N

IMPROVED PORTABLE
CONTRACTORS' ENGINE,

With Patent Feed Water Heater, and Patent Steam Blast Tube Cleaner.

The Patent Feed Water Heater raises the feed to near the boiling point, saving much fuel and wear and tear of Fire Boxes and Tubes.

The Patent Steam Blast Tube Cleaner entirely dispenses with the use of Tube Brushes. By the simple movement of a handle, a powerful blast of steam is introduced into the chimney, instantly withdrawing from the Tubes all soot and other obstructions.

The Material and Workmanship employed in the construction of these Engines is of the very first class, and their finish is unrivalled.

IMPROVED PORTABLE CONTRACTORS' ENGINE,—*continued.*

Nominal Horse Power.	Diameter of Cylinder in inches.	Length of Stroke in inches.	Speed in Revolutions per minute.	Diameter of Fly-Wheel in inches.	Diameter of Crank Shaft in inches.	Weight without case in cwts.	Price.	Nominal Horse-Power.
2½	5¼	8	180	42	2	30	£105	2½
3	5¾	9	150	48	2¼	42	125	3
3½	6¼	9	150	48	2¼	45	140	3½
4	6¾	10	130	56	2½	57	150	4
5	7½	12	130	60	2¾	62	165	5
6	8½	12	120	60	2¾	72	180	6
7	9	14	110	66	3	85	195	7
8	9¾	14	110	66	3	88	210	8
9	10¼	14	110	66	3¼	93	225	9
10	10¾	14	110	66	3¼	97	240	10
12	12	14	110	66	3½	110	280	12

All Engines above 8½ Horse Power are supplied with the patent Feed Water Heater. Link Motion Reversing Gear, £15 extra.

F. W. REYNOLDS, & Co., Engineers, 73, Southwark Street: And The Grove, London, S.E.

F. W. REYNOLDS & CO.S

New Patent "Imperial" Combination Self-Feeding Saw Bench. Specially Designed for Hand-Power, Price, Complete, £22.

The " Imperial " Circular Saw Bench, £12 10s.

With Rising and Falling Spindle, £14 10s.

The special object in the construction of this Machine has been to secure the greatest speed and power with the minimum expenditure of labour, in other words, to produce a machine capable of performing general work, and which could be driven easily by a man or lad of average strength. This result has been attained in F.W. Reynolds and Co.'s " Imperial " Combined Circular and Band Sawing Machine, by the Circular Saw of which timber up to 4½ inches thick can be cut, and by the Band Saw of which 5½ inches can be cut. Two Double Cog Wheels to change are provided, whereby the rate of Feed Motion is varied according to the thickness of timber in cut. There is a Fence Plate parallel to the Saw, fitted with set screws at back, whereby it can be regulated and adjusted ; it may also be set at any angle for Bevel Cutting. A groove in the Table serves to guide a Cross Cutting and Mitreing Fence by which this work is done with great accuracy and expedition. For Tenoning or Grooving the Feeding Apparatus must be released and shifted over ; the nut being taken off end of Feed Shaft, and the small set screw in the Wheel released, the Shaft may be drawn out of the way.

F. W. REYNOLDS, & Co., Engineers, 73, Southwark Street :
And The Grove, London, S.E.

MORTISING MACHINES.

The "MONARCH"

Patent Mortising Machine.

Price complete, with
chisels and core
divers, £12.

The "ROYAL" Patent Combined Mortising, Tenoning,
and Boring Machine.

Price complete, with chisels,
augurs, &c. £21.

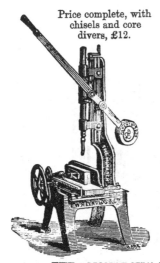

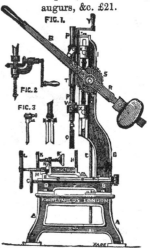

THE "MONARCH" MORTISING MACHINE.

Messrs. F. W. REYNOLDS & Co., of Southwark Street, have just
introduced a new Mortising Machine which they call the "Monarch,"
combining the most recent improvements with several advantages
peculiar to itself. The raising and lowering of the brackets in which
the chisel-carrying spindle works, is effected by a new arrangement
holding them quite firm while working in different depths of stuff;
the spindle consequently is not liable to fall suddenly, as it does in
some machines, damaging the chisel. Additional power is gained by
the direct action of the lever handle, and the friction is reduced to a
minimum. The machine is mounted on an extra strong stand, and
will, we should think, soon come into general use among the trade.
The same firm have, we understand, made several improvements in
their Imperial hand power Saw Bench, and they have a new patent
combined circular and band sawing machine nearly ready for
delivery.—*From the "Building News," Dec.* 26, 1874.

*F. W. REYNOLDS, & Co., Engineers, 73, Southwark Street :
And The Grove, London, S.E.*

THE "ECLIPSE" NEW PATENT MORTISING MACHINE SECURES PERFECT MORTISES.

The novelty of this Machine (" THE ECLIPSE "), consists in the method of connecting the hand-lever to the part which carries the spindle of the chisel. In place of the usual rack and cog wheel, it has an articulated elbow, or toggle movement, the action of which not only produces the requisite vertical motion of the chisel, but gives it an increased power as it penetrates deeper and deeper into the material being worked. The regularity of the depth of mortices not required to pass through the wood is secured by making the part adjustable which carries the chief spindle of the fulcrum of the lever. This Machine is made of great strength throughout, and is thoroughly well fitted and finished.

Price Complete,£16 16s.

BAND SAW TABLES, TENONING MACHINES, BORING MACHINES.

General Joiners, and all kinds of Wood Working and other Builders' Machinery, in great Variety.

F. W. REYNOLDS, & Co., Engineers, 73, Southwark Street: And The Grove, London, S.E.

F. W. REYNOLDS & CO.,

73, Southwark Street, and The Grove, LONDON, S.E.,

WHOLESALE AGENTS FOR

ROBINS' LONDON PORTLAND CEMENT.

THE various experiments made with this Cement by the Engineers of the Government Board of Works prove it to be 20 per cent. above the Government test for its strength. IT WAS AWARDED THE ONLY PRIZE MEDAL of the Great Exhibition of 1851, and for many years has enjoyed the preference in the Russian and German trade ; has been most extensively used in the construction of the Main Drainage of London, also for Docks, Waterworks, Foundations for Bridges, etc., both in this country and abroad. Numerous testimonials from Builders who have used it for Cementing Buildings during the past twenty years.

PRICES.

Per CASK, 400 lbs., 12s. 6d. 3s. allowed when empty cask received, carriage paid.

Per SACK, 200 lbs., 6s. 1s. 6d. allowed when empty sack received, carriage paid, within 28 days.

LONDON OFFICES AND WAREHOUSES :

73, SOUTHWARK STREET, & THE GROVE, LONDON, S.E.

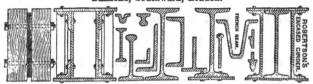

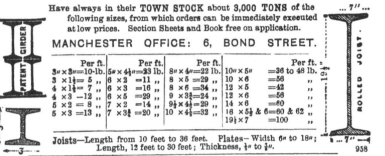

upwards. These standards are fastened at bottom by a projecting shoe or foot to stumps driven in the ground, and kept perpendicular by angle iron stays or braces.

Ordinary rough inch boards are employed in preference to wood or iron panels, between which the Concrete is placed in a soft state ; simple iron cramps, also connected by wall gauges in a similar manner to the standards, keep the boards in position, aided by the fixed guides or standards.

Angle irons for forming the external angles, &c., of walls, side irons for cross walls, joint irons for chimney breasts, piers and counter-forts, and scaffold brackets to fix to the wall guides are of simple construction and easily adjusted.

The inner angles of walls are simply formed by the fixed standards.

The only objections hitherto raised against the use of common boards in place of tha patented cumbersome panels are, that the effect of the wet Concrete on one side only would cause the boards to warp or twist, and that the waste by depreciation would more than equal the cost of patented panels, but these assertions are not borne out in practice.

The boards can be reversed each time they are moved, and this prevents the warping or distortion, and the value of time saved in fixing the new apparatus, added to the cost of maintaining in repair the usual patented panels, amount to considerably more than the loss and depreciation in value of boarding, the entire cost of which amounts to but little more than £3 for each 100 feet lineal of wall buildings at one time.

For plain straight walling the cost of "POTTER'S PATENT APPLIANCES" would be without scaffold brackets, angle irons, joint irons, &c., about 8/- per foot run of walling required to be constructed at the same time, and increasing in cost with the more complicated character of the buildings, which would necessitate the accessories previously described.

The appliances can be seen in use, and further particulars, estimates, &c., had on application to F. W. REYNOLDS & Co.,

---

F. W. REYNOLDS & CO.,

ENGINEERS,

73, SOUTHWARK STREET, & THE GROVE,

LONDON, S.E.,

SOLE AGENTS.

Printed in the United States
By Bookmasters